Valery Yundin

Massive Loop Corrections for Collider Physics

Valery Yundin

Massive Loop Corrections for Collider Physics

Tensor Integrals in Field Theory

Südwestdeutscher Verlag für Hochschulschriften

Impressum/Imprint (nur für Deutschland/only for Germany)
Bibliografische Information der Deutschen Nationalbibliothek: Die Deutsche Nationalbibliothek verzeichnet diese Publikation in der Deutschen Nationalbibliografie; detaillierte bibliografische Daten sind im Internet über http://dnb.d-nb.de abrufbar.
Alle in diesem Buch genannten Marken und Produktnamen unterliegen warenzeichen-, marken- oder patentrechtlichem Schutz bzw. sind Warenzeichen oder eingetragene Warenzeichen der jeweiligen Inhaber. Die Wiedergabe von Marken, Produktnamen, Gebrauchsnamen, Handelsnamen, Warenbezeichnungen u.s.w. in diesem Werk berechtigt auch ohne besondere Kennzeichnung nicht zu der Annahme, dass solche Namen im Sinne der Warenzeichen- und Markenschutzgesetzgebung als frei zu betrachten wären und daher von jedermann benutzt werden dürften.

Coverbild: www.ingimage.com

Verlag: Südwestdeutscher Verlag für Hochschulschriften GmbH & Co. KG
Heinrich-Böcking-Str. 6-8, 66121 Saarbrücken, Deutschland
Telefon +49 681 37 20 271-1, Telefax +49 681 37 20 271-0
Email: info@svh-verlag.de

Herstellung in Deutschland:
Schaltungsdienst Lange o.H.G., Berlin
Books on Demand GmbH, Norderstedt
Reha GmbH, Saarbrücken
Amazon Distribution GmbH, Leipzig
ISBN: 978-3-8381-3210-5

Imprint (only for USA, GB)
Bibliographic information published by the Deutsche Nationalbibliothek: The Deutsche Nationalbibliothek lists this publication in the Deutsche Nationalbibliografie; detailed bibliographic data are available in the Internet at http://dnb.d-nb.de.
Any brand names and product names mentioned in this book are subject to trademark, brand or patent protection and are trademarks or registered trademarks of their respective holders. The use of brand names, product names, common names, trade names, product descriptions etc. even without a particular marking in this works is in no way to be construed to mean that such names may be regarded as unrestricted in respect of trademark and brand protection legislation and could thus be used by anyone.

Cover image: www.ingimage.com

Publisher: Südwestdeutscher Verlag für Hochschulschriften GmbH & Co. KG
Heinrich-Böcking-Str. 6-8, 66121 Saarbrücken, Germany
Phone +49 681 37 20 271-1, Fax +49 681 37 20 271-0
Email: info@svh-verlag.de

Printed in the U.S.A.
Printed in the U.K. by (see last page)
ISBN: 978-3-8381-3210-5

Copyright © 2012 by the author and Südwestdeutscher Verlag für Hochschulschriften GmbH & Co. KG and licensors
All rights reserved. Saarbrücken 2012

Contents

1. **Introduction**   1

2. **Theoretical background**   4
   - 2.1. Basics of perturbative QFT 4
   - 2.2. QED Lagrangian and Feynman rules 10
   - 2.3. Dimensional regularization 12
   - 2.4. Renormalization 15
   - 2.5. Cancellation of IR divergences 17

3. **One-loop tensor reduction**   19
   - 3.1. Definitions and notation 19
     - 3.1.1. Dimensionally regulated tensor integrals 19
     - 3.1.2. Kinematic determinants 21
   - 3.2. Passarino-Veltman reduction 22
   - 3.3. Dimensional recurrence method 26
     - 3.3.1. Scalar integrals in shifted dimension 27
     - 3.3.2. Reduction of 5-point functions 28
     - 3.3.3. Reduction of 4-point functions 34
     - 3.3.4. Lower point functions 36

4. **Numerical code PJFry**   38
   - 4.1. Program description 39
   - 4.2. Evaluation of scalar integrals 41
     - 4.2.1. Evaluation of UV poles 43
     - 4.2.2. Convergence acceleration 46
   - 4.3. Interface and usage examples 47

		4.3.1. Cache system	50
		4.3.2. Mathematica interface	53
	4.4.	Fully contracted tensor test	60
		4.4.1. Massless configuration	63
		4.4.2. Massive configuration	63
	4.5.	Five gluon amplitude test	65
		4.5.1. Calculation method	66
		4.5.2. Accuracy plots	69

5. QED virtual corrections to the process $e^+e^- \to \mu^+\mu^-\gamma$ — **71**

    5.1. Notation and conventions 73
    5.2. Computation method 74
        5.2.1. Born amplitude 75
        5.2.2. Loop amplitude 76
    5.3. Renormalization and scheme dependence 79
    5.4. Pole structure and μ_R-dependence 80
    5.5. Cross-checks 82
    5.6. Numerical results 83

6. Summary and outlook — **93**

A. Algebra of signed minors — **96**

B. Scalar integral recurrence relations — **99**

C. On-shell two point functions — **103**

D. Diagrams for $e^+e^- \to \mu^+\mu^-\gamma$ — **106**

Bibliography — **110**

Acknowledgements — **130**

1. Introduction

The Standard Model (SM) is a theory of elementary particle interactions. In the last 50 years the Standard Model has been verified in a wide range of experiments over many orders of magnitude of the energy spectrum. The discovery of top quark and tau neutrino completed the picture in the strong and electroweak sectors of the SM in total agreement with theoretical predictions.

Despite its great success, the Standard Model can not be regarded as a complete theory of fundamental interactions. Leaving out the gravitation and dark energy, which are not described by the SM, we have more immediate problems of explaining the electroweak (EW) symmetry breaking mechanism and neutrino oscillations. A wide spectrum of theories trying to deal with these problems includes new types of elementary particles.

The generally accepted solution to the origin of particle masses and therefore the EW symmetry breaking is the Higgs mechanism. It requires the existence of a neutral scalar particle, the Higgs boson. Yet unobserved, the Higgs boson is a part of the SM, which would make little sense without it.

The experimental observation of the Higgs boson or possibly other new particles is the main priority of current high energy collider experiments at Tevatron and LHC. The production of new particles is accompanied by the large SM backgrounds which need to be accurately predicted in order to make the discovery of new physics possible. At current collision energies these backgrounds routinely include many particle final states with massless and massive particles.

Such precision tests of the Standard Model at existing and future high energy colliders require the knowledge of the theory parameters to very high accuracy. The measurements of fine structure constant α and anomalous magnetic moment of the muon α_μ at high luminosity e^+e^- colliders (VEPP-

2M, DAΦNE, BEPC, PEP-II and KEKB) are reaching an unprecedented accuracy, which has to be matched by precision calculations on the theory side [1].

Due to the complexity of quantum field theories, the predictions for high energy processes are calculated as terms of the perturbative expansion in the interaction constant. The calculations in the leading order (LO) of this expansion nowadays can be performed in a completely automated fashion [4, 46, 131, 133, 148]. Unfortunately the accuracy of the LO is not sufficient for confirming or disproving the SM and its extensions. This is especially true for perturbative Quantum Chromodynamics (QCD) for which next to leading order (NLO) calculations are needed in order to reliably estimate the normalization of cross-sections [55].

Over the last decade there has been substantial progress in multileg NLO calculations of many $2 \to 3$ and some $2 \to 4$ processes by different groups [23, 33, 34, 36, 42, 47, 48, 49, 72, 78, 79, 84, 85, 91, 100, 108, 135, 136, 137, 138]. Several frameworks for NLO calculations have been developed although the level of automation has not reached that of LO cross-sections [16, 22, 35, 64, 112, 118, 145].

All these one-loop multileg calculations have been done with one of the two competing methods. One is a family of methods generally refered to as *unitarity approaches*, which is based on the analytic properties of unitarity and factorization of one-loop amplitudes [29, 31, 50, 89, 92, 144]. It is very efficient for calculations of processes with many gauge bosons and other massless particles. The other one is the traditional *Feynman diagram* approach which is competitive for the calculations involving massive intermediate states (e.g. t-quark, massive gauge bosons, etc.).

One essential component of any Feynman diagram NLO calculation is the evaluation of tensor loop integrals. The standard Passarino-Veltman approach to tensor integral evaluation fails for multileg processes due to the small Gram determinant problem. Several alternative reduction schemes have been proposed. At the moment there is no publicly available code implementing those algorithms for arbitrary kinematics with internal massive particles.

Outline

In the present work we discuss the problem of Gram determinants in the evaluation of the one-loop tensor integrals with focus on the dimension recurrence method, its numerical implementation PJFry and applications to photon associated muon pair production at one loop level.

The thesis is organized as follows.

In the first chapter we briefly remind the reader of the basic concepts of perturbative quantum field theory, which are extensively used throughout the rest of the document. We begin by outlining the steps leading from the Lagrangian of the theory to Feynman rules. Which is followed by the discussion of the origins of divergences in the loop integrals and their regularization and subtraction.

In the second chapter we discuss the methods and the problems of evaluation of one loop tensor integrals in dimensional regularization. We start with the simple Passarino-Veltman reduction and demonstrate the appearance of inverse Gram determinants, which could cause numerical instabilities. Then we show how Gram determinants can be avoided by switching to another integral basis. In the rest of the chapter the reduction scheme based on the dimensional recurrence relations is explained in detail.

In chapter 3 we present the numerical library for tensor integral evaluation based on the methods of chapter 2. Starting from a general description we proceed to the methods of basis integral evaluation and algorithm selection. The chapter is concluded by several accuracy tests for massless and massive kinematic configurations.

The last chapter is devoted to the calculation of the one-loop virtual corrections to muon pair production with hard photon emission. The computation method is explained, followed by the discussion of renormalization and pole structure. In the numerical results we present differential distributions for realistic kinematics of the KLOE detector at DAΦNE in Frascati and the BaBar detector at SLAC.

2. Theoretical background

2.1. Basics of perturbative QFT

Let us briefly overview basic concepts of perturbative quantum field theory taking self-interacting scalar field $\phi(x)$ as an example.

The defining object of the theory is an action $S = \int dt\, L$, where L is the Lagrangian. We write it as the space-time integral of a Lagrangian density which we separate into free field part $\mathcal{L}_0$ and interaction part $\mathcal{L}_{\text{int}}$

$$S = \int d^4x\, \mathcal{L}, \qquad \mathcal{L} = \mathcal{L}_0 + \mathcal{L}_{\text{int}}, \tag{2.1}$$

$$\mathcal{L}_0 = \frac{1}{2}\partial^\mu \phi\, \partial_\mu \phi - \frac{1}{2}m^2\phi^2, \qquad \mathcal{L}_{\text{int}} = \frac{1}{6}\lambda\phi^3. \tag{2.2}$$

The field $\phi(x)$ obeys the equation of motion

$$(\partial^2 + m^2)\phi = \frac{1}{2}\lambda\phi^2. \tag{2.3}$$

The scattering amplitude of n initial particles with momenta p_i into n' final particles with momenta p'_i can be written with the help of the Lehmann-Symanzik-Zimmermann reduction formula (LSZ formula) for scalar fields

$$\begin{aligned}\langle F_{n'}|I_n\rangle = i^{n+n'} \int \cdots \int & d^4x_1\, e^{ip_1 x_1}(\partial_1^2 + m_1^2)\, d^4x_2\, e^{ip_2 x_2}(\partial_2^2 + m_2^2) \ldots \\ & d^4x'_1\, e^{-ip'_1 x'_1}(\partial_{1'}^2 + m_{1'}^2)\, d^4x'_2\, e^{-ip'_2 x'_2}(\partial_{2'}^2 + m_{2'}^2) \ldots \\ & \times \langle 0|T\phi(x_1)\phi(x_2)\ldots\phi(x_n)\phi(x'_1)\phi(x'_2)\ldots\phi(x'_{n'})|0\rangle\end{aligned} \tag{2.4}$$

where $\langle 0|T\phi(x_1)\ldots\phi(x'_1)\ldots|0\rangle$ is the vacuum expectation value of a time-ordered product of field operators, also known as correlation function or Green function. The interacting field $\phi(x)$ satisfies a normalization condition

$$\langle 0|\phi(x)|0\rangle = 0, \qquad \langle p|\phi(x)|0\rangle = e^{-ipx}. \tag{2.5}$$

Introducing an auxiliary source field J we define the functional integral of the theory

$$Z(J) \equiv \langle 0|0\rangle_J = \int \mathcal{D}\phi \, e^{i\int d^4x [\mathcal{L}_0 + \mathcal{L}_{\text{int}} + J\phi]} \tag{2.6}$$

where the integration is performed over the space of field configurations

$$\mathcal{D}\phi \propto \prod_x \phi(x) \tag{2.7}$$

and the measure is assumed to be chosen so the integral is normalized as $Z(0) = 1$.

Using the functional integral notation the n-point correlation function can be written as

$$\langle 0|T\phi(x_1)\ldots\phi(x_n)|0\rangle = \frac{\int \mathcal{D}\phi \, \phi(x_1)\ldots\phi(x_n) e^{iS}}{\int \mathcal{D}\phi \, e^{iS}} \tag{2.8}$$

or as functional derivative of the generating functional $Z(J)$

$$\langle 0|T\phi(x_1)\ldots\phi(x_n)|0\rangle = (-i)^n \frac{\delta^n Z(J)}{\delta J(x_1)\ldots\delta J(x_n)}\bigg|_{J=0} \tag{2.9}$$

The $Z(J)$ can be split into the free field part and interaction part

$$Z(J) = \exp\left[i\int d^4x \, \mathcal{L}_{\text{int}}\left(\frac{1}{i}\frac{\delta}{\delta J(x)}\right)\right] Z_0(J) \tag{2.10}$$

where $Z_0(J)$ is the free field part, which can be expressed in terms of known 2-point Green functions $\Delta(x - x')$

$$Z_0(J) = \exp\left[\frac{i}{2} \int d^4x \, d^4x' \, J(x)\Delta(x - x')J(x')\right] \tag{2.11}$$

$$\Delta(x - x') = \int \frac{d^4q}{(2\pi)^4} \frac{e^{iq(x-x')}}{q^2 - m^2 + i\epsilon} \tag{2.12}$$

By substituting $\mathcal{L}_{\text{int}}$ and expanding the exponent in (2.10) we can obtain a power series expansion of $Z(J)$ in terms of interaction constant λ. Combining it with expanded $Z_0(J)$ we get

$$Z(J) = \sum_{V=0}^{\infty} \frac{1}{V!} \left[\frac{i\lambda}{6} \int d^4x \left(\frac{1}{i}\frac{\delta}{\delta J(x)}\right)^3\right]^V \times \sum_{P=0}^{\infty} \frac{1}{P!} \left[\frac{i}{2} \int d^4y \, d^4x \, J(y)\Delta(y-z)J(z)\right]^P \tag{2.13}$$

Putting the above expression (2.9) will give us a *perturbative expansion* of the correlation function, which we can later insert in LSZ formula in order to get scattering amplitudes.

The direct use of (2.9) in this way would require evaluating lots and lots of functional derivatives of $Z_0(J)$. This computation can be organized by introducing Feynman diagrams. In these diagrams we assign

- lines to propagators $\Delta(x_1 - x_2)$
- vertices to $i\lambda \int d^4x$
- endpoints to $i \int d^4x \, J(x)$

In each diagram V is the number of vertices, P is the number of propagators and $E = 2P - 3V$ is the number of endpoints (legs). The $Z(J)$ is then given by the sum of all possible diagrams D

$$Z(J) = \sum D \tag{2.14}$$

The general diagram D can be written as a product of connected diagrams C_I

$$D = \prod_I \frac{1}{n_I!} (C_I)^{n_I} \qquad (2.15)$$

where n_I is the number of identical diagrams C_I and the factorial corresponds to the symmetry factor. That allows us to rewrite the path integral as exponential of the sum of all connected diagrams

$$Z(J) = \sum D = \sum \prod_I \frac{1}{n_I!}(C_I)^{n_I} = \prod_I \sum_{n_I=0}^{\infty} \frac{1}{n_I!}(C_I)^{n_I} = \exp\left(\sum_I C_I\right) \qquad (2.16)$$

In order to satisfy the normalization condition $Z(0) = 1$ we should drop the vacuum diagram (with no endpoints), which gives us

$$Z(J) = \exp\left[iW(J)\right] \qquad \text{with} \qquad iW(J) \equiv \sum_{I \neq \{0\}} C_I$$

where $I \neq \{0\}$ means that we omit vacuum diagrams and $W(0) = 0$.

In this form $Z(J)$ still contains disconnected products of connected diagrams. By substituting them into the correlation function (2.9) and using the LSZ formula one can show that these terms correspond to several independent scattering events happening at the same time. This is usually not what we want to calculate. Therefore we define a connected correlation function, which will describe a single multiparticle scattering event after being inserted in the LSZ formula

$$\langle 0|T\phi(x_1)\ldots\phi(x_n)|0\rangle_C = (-i)^{n-1} \left.\frac{\delta^n W(J)}{\delta J(x_1)\ldots\delta J(x_n)}\right|_{J=0}. \qquad (2.17)$$

For example let us calculate the $2 \to 2$ particle scattering amplitude. First we need the 4-point connected correlation function $\langle 0|T\phi(x_1)\phi(x_2)\phi(x_1')\phi(x_2')|0\rangle_C$. The lowest non-zero order in λ is contributed by the diagrams with $V = 2$ and $P = 5$, shown in Fig. 2.1. These are so-called *tree* diagrams (no closed loops).

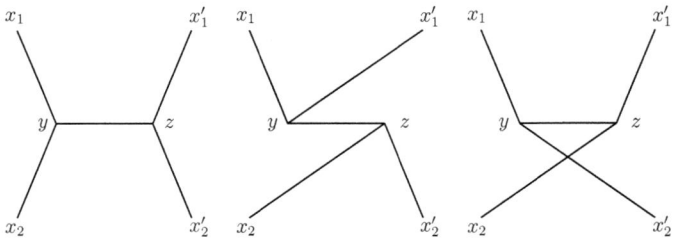

Figure 2.1.: Tree diagrams for $2 \to 2$ scattering in ϕ^3

Using these diagrams we can write down the correlation function

$$\begin{aligned}
\langle 0|T\phi(x_1)\phi(x_2)\phi(x_1')\phi(x_2')|0\rangle_C = &-i(i\lambda)^2 \int d^4y\, d^4z\, \Delta(y-z) \\
&\times \big[\Delta(x_1-y)\Delta(x_2-y)\Delta(x_1'-z)\Delta(x_2'-z) \\
&+ \Delta(x_1-y)\Delta(x_1'-y)\Delta(x_2-z)\Delta(x_2'-z) \\
&+ \Delta(x_1-y)\Delta(x_2'-y)\Delta(x_2-z)\Delta(x_1'-z)\big] + O(\lambda^4)
\end{aligned} \quad (2.18)$$

We can now substitute this to the LSZ formula and use the 2-point Green function (2.12). After working out the details we arrive at the momentum space representation of the $2 \to 2$ scattering amplitude

$$\begin{aligned}
\langle 1'2'|12\rangle = &\; i(2\pi)^4\delta^4(p_1+p_2-p_1'-p_2') \\
&\times \lambda^2 \left[\frac{1}{(p_1+p_2)^2-m^2} + \frac{1}{(p_1-p_1')^2-m^2} + \frac{1}{(p_1-p_2')^2-m^2}\right] + O(\lambda^4)
\end{aligned} \quad (2.19)$$

The first part with the momenta conservation delta function is common to all scattering amplitudes. It is convenient to consider only the second part, which is process specific and called *matrix element M*

$$\langle F_{n'}|I_n\rangle = (2\pi)^4\delta(p_{\text{in}}-p_{\text{out}})iM \quad (2.20)$$

The diagrammatic calculation of matrix elements can be elegantly described by the following algorithm

1. Each particle type has its own line type. Draw *external* lines for each incoming and each outgoing particle.

2. Connect all external lines leaving one end free and using lines and vertices available in the theory. In this way draw all possible *fully connected* diagrams that are topologically nonequivalent.

3. On each incoming line draw an arrow pointing towards the connected vertex. On each outgoing line draw an arrow pointing away from the connected vertex. On each internal line draw an arrow with arbitrary direction.

4. Assign each line its own momentum. The momentum of external line is the momentum of the corresponding particle. Thinking as if momenta are flowing along the lines in the direction of the arrows, assign momenta to internal lines using momenta conservation at the vertices.

5. The *value* of the diagram is obtained by replacing vertices and lines according to theory-specific Feynman rules.

6. The value of each diagram is divided by its symmetry factor which is equal to the order of the permutation group of internal lines and vertices leaving the diagram unchanged.

7. For tree diagrams all momenta will be fixed. For diagrams with L closed loops integrate over each free momentum k_i with measure $d^4 k_i/(2\pi)^4$.

8. The matrix element M is given by the sum over *values* of all these diagrams.

The Feynman rules can be derived from the interaction part $\mathcal{L}_{\text{int}}$ of the Lagrangian of the theory. In scalar ϕ^3 theory we have one 3-vertex and the following rules:

- external line gives 1

- internal line with momentum p gives $-i/(p^2 - m^2 + i\epsilon)$
- vertex gives $i\lambda$

One can easily verify that applying these rules gives (2.19) for $2 \to 2$ scattering amplitude.

The scattering amplitude (and its matrix element) is the main component of the differential cross section, which enters definitions of physical observables. The fully differential cross section of $2 \to n'$ particles scattering is

$$d\sigma = \frac{1}{4|\mathbf{p}_1|_{\text{CMS}}\sqrt{s}}|M|^2\, d\Phi_{1+2\to n'} \qquad (2.21)$$

where $s = (p_1+p_2)^2$, $|\mathbf{p}_1|_{\text{CMS}}$ is the absolute 3-momentum in the center-of-mass frame of incoming particles and $d\Phi_{1+2\to n'}$ is the n'-body Lorentz-invariant phase-space measure.

2.2. QED Lagrangian and Feynman rules

The Quantum Electrodynamics (QED) is the theory describing interactions of charged spin $\frac{1}{2}$ particles with the electromagnetic field. Its Lagrangian density is

$$\mathcal{L}_{\text{QED}} = -\frac{1}{4}F_{\mu\nu}F^{\mu\nu} + i\bar{\psi}\partial_\mu\gamma^\mu\psi - m\bar{\psi}\psi - e\bar{\psi}\gamma^\mu\psi A_\mu \qquad (2.22)$$

The calculation of the correlation functions is a bit more involved due to presence of vector and spinor fields. We list here the result of this calculation and refer the reader to textbooks [147, 157] for technical details. The Feynman rules in the momentum space for computing contributions to matrix elements in QED read

- $u_s(p)$ incoming fermion
- $\bar{u}_s(p)$ outgoing fermion
- $\bar{v}_s(p)$ incoming anti-fermion

- $v_s(p)$ outgoing anti-fermion
- $\epsilon^\mu_{(\lambda)}(p)$ incoming photon
- $\epsilon^{\mu*}_{(\lambda)}(p)$ outgoing photon
- $\dfrac{i(q_\mu \gamma^\mu + m)}{q^2 - m^2 + i\epsilon}$ fermion propagator
- $\dfrac{-ig^{\mu\nu}}{q^2 + i\epsilon}$ photon propagator
- $-ieQ_e \gamma^\mu$ photon-fermion vertex

In addition to momentum flow each fermion line gets assigned a *fermion flow*. The spinors and gamma matrices do not commute and the value of the diagram along the spinor line should be written left to right *against* the fermion flow. Fermion line loops get an additional factor of -1 and the trace operation has to be applied to gamma matrices along the loop.

The polarization spinors u and v satisfy the Dirac equation

$$(p_\mu \gamma^\mu - m)u_s(p) = 0 \qquad \bar{u}_s(p)(p_\mu \gamma^\mu - m) = 0 \qquad (2.23)$$
$$(p_\mu \gamma^\mu + m)v_s(p) = 0 \qquad \bar{v}_s(p)(p_\mu \gamma^\mu + m) = 0 \qquad (2.24)$$

and the completeness relations

$$\sum_s u_s(p)\bar{u}_s(p) = p_\mu \gamma^\mu + m \qquad \sum_s v_s(p)\bar{v}_s(p) = p_\mu \gamma^\mu - m \qquad (2.25)$$

The Dirac gamma matrices obey the Clifford algebra generated by the anticommutation relation

$$\{\gamma^\mu, \gamma^\nu\} = 2g^{\mu\nu}\mathbf{1} \qquad (2.26)$$

which results in a number of useful contraction identities

$$\begin{aligned}
\gamma^\mu \gamma^\nu \gamma_\mu &= -(g^\mu_{\ \mu} - 2)\gamma^\nu \\
\gamma^\mu \gamma^\nu \gamma^\lambda \gamma_\mu &= 4g^{\nu\lambda} + (g^\mu_{\ \mu} - 4)\gamma^\nu \gamma^\lambda \\
\gamma^\mu \gamma^\nu \gamma^\lambda \gamma^\sigma \gamma_\mu &= -2\gamma^\sigma \gamma^\lambda \gamma^\nu - (g^\mu_{\ \mu} - 4)\gamma^\nu \gamma^\lambda \gamma^\sigma \\
\gamma^\mu \gamma^\nu \gamma^\lambda \gamma^\sigma \gamma^\rho \gamma_\mu &= 2\gamma^\sigma \gamma^\lambda \gamma^\nu \gamma^\rho + (g^\mu_{\ \mu} - 4)\gamma^\nu \gamma^\lambda \gamma^\sigma \gamma^\rho + 2\gamma^\rho \gamma^\nu \gamma^\lambda \gamma^\sigma
\end{aligned} \quad (2.27)$$

$\ldots$

The leading order contributions to matrix elements are described by tree level diagrams with all internal momenta fixed by the momentum conservation. On the other hand, contributions beyond leading order contain diagrams with closed loops which require additional integrations over the undetermined loop momenta.

Our ability to calculate the perturbative expansion for the scattering amplitudes is mainly limited by the complexity of these loop integrations, which grow very quickly with the increasing number of loops and/or external states. In fact in the case of multiparticle processes even the first one-loop correction presents enough technical difficulties which are not completely solved up to date.

2.3. Dimensional regularization

The first problem one encounters when trying to evaluate loop integrals is the fact that they are often divergent (have infinite value). The physical implications of this will be discussed in the next section, while now we will try to deal with it from a purely mathematical point of view.

The example of a divergent diagram is the first loop correction to the fermion propagator. The corresponding diagram and its value are shown below. Note that we consider it here as a part of a bigger diagram, therefore polarization spinors are not included.

$$\text{[diagram]} = \frac{i(\slashed{p}+m)}{p^2-m^2}[-i\Sigma_2(p)]\frac{i(\slashed{p}+m)}{p^2-m^2} \qquad (2.28)$$

where

$$-i\Sigma_2(p) = (-ie)^2 \int \frac{d^4k}{(2\pi)^4} \gamma^\mu \frac{i(\slashed{k}+m)}{k^2-m^2}\gamma^\nu \frac{-ig_{\mu\nu}}{(p-k)^2} \qquad (2.29)$$

The integration in (2.29) has logarithmic divergence $\sim \log k^2$ at large loop momentum $k^2 \to \infty$. Such divergence occuring at large loop momentum is called ultraviolet (UV). If one tries to calculate the vertex correction he will encounter another kind of divergence, which is called infrared (IR) and appears at low $k^2 \to 0$.

We have to make integral (2.29) well defined and it can be done by introducing a regulator. The infrared and ultraviolet divergences have different origins and in one principle can use different regulators, for example a small photon mass for the IR divergence and the Pauli-Villars procedure for the UV divergence. Besides some pedagogical value there is little reason to do so in practical calculations. We will use dimensional regularization [12, 45, 59, 161] which handles both kinds of divergence in an efficient and uniform fashion.

In the dimensional regularization we formally extend the loop momentum k into a $d = 4 - 2\epsilon$ dimensional space which makes the integral parametrically dependent on d. The value of d is chosen to make the d-dimensional integration well-defined and can be non-integer or even complex. After performing the integration we should analytically continue the result to $d = 4$. The divergences will manifest themselves as poles in the Laurent expansion around $\epsilon = 0$.

Applying this procedure to the integration in (2.29) and omitting technical details we get

$$-i\Sigma_2(p) = (-ie)^2 \mu^{2\epsilon} \int \frac{d^d k}{(2\pi)^4} \gamma^\mu \frac{i(\slashed{k}+m)}{k^2-m^2} \gamma^\nu \frac{-ig_{\mu\nu}}{(p-k)^2} \quad (2.30)$$

$$= \frac{1}{\epsilon}(-ie)^2 \frac{2i}{(4\pi)^2}\left(\frac{4\pi\mu^2}{m^2}\right)^\epsilon \Gamma(1+\epsilon)(4m-\slashed{p}) + \text{finite} + O(\epsilon) \quad (2.31)$$

This result demonstrates the main features of the dimensional regularization approach

- ultraviolet and/or infrared singularities are explicitly separated into poles in the dimensional regulator ϵ.

- the finite part contains physical information at the one loop level, and further terms in the ϵ-expansion contribute to higher order corrections.

- a new scale with the dimension of mass μ was introduced in order to preserve the overall dimensionality of the diagram.

Over the last 50 years many methods for the analytic evaluation of dimensionally regulated Feynman integrals have been developed. A good overview of contemporary state of art techniques is available in [156].

There is a freedom in the definition of dimensional regularization connected with the dimensionality of Dirac gamma matrices, the associated Clifford algebra and the dimensionality of vector fields and their polarization vectors. This results in the coexistence of several variants of dimensional regularization usually called *schemes* [153, 154, 155]:

- Conventional Dimensional Regularization (CDR) – all vector fields are uniformly continued to d dimensions. The numerator gamma algebra is done in d dimensions

- 't Hooft Veltman regularization (HV) – internal vector fields are treated as d-dimensional and external ones are treated as strictly 4-dimensional. The numerator gamma algebra is done in d dimensions.

- **Four Dimensional Helicity scheme (FDH)** – internal vector fields are treated as d-dimensional and external ones are treated as strictly 4-dimensional. The numerator gamma algebra is done in 4 dimensions.

All schemes have in common that the loop momentum is always continued to d dimensions. At one loop level these schemes are equivalent and the results of a calculation can be converted from one scheme to another one with simple transitional formulae [57, 126, 127]. The formally d-dimensional space is in fact an infinite-dimensional vector space with some d-dimensional properties. For a proper definition of this space and operations in it see [60, 175].

2.4. Renormalization

To start dealing with singularities we should first remember that the LSZ formula is only valid if the interacting fields satisfy the normalization of the free fields (2.5). The same conditions for spinor and vector fields are

$$\langle 0|\psi(x)|0\rangle = 0, \qquad \langle p,s,-|\bar{\psi}(x)|0\rangle = v_s(p)\,e^{-ipx}, \qquad (2.32a)$$
$$\langle 0|A^\mu(x)|0\rangle = 0, \qquad \langle p,\lambda|A^\mu(x)|0\rangle = \epsilon^\mu_{(\lambda)}(p)\,e^{ipx}. \qquad (2.32b)$$

These relations generally do not hold for the interacting fields, because they receive corrections from loop-diagrams. We can fix them by shifting and rescaling parameters in the original Lagrangian (2.22)

$$\psi \to \sqrt{Z_2}\psi, \quad A^\mu \to \sqrt{Z_3}A^\mu, \quad m \to (1+Z_m/Z_2)m, \quad e \to Z_1/(Z_2\sqrt{Z_3})e, \tag{2.33}$$

$$\mathcal{L}_{\text{QED}} = -\frac{1}{4}Z_3 F_{\mu\nu}F^{\mu\nu} + Z_2\bar{\psi}(i\partial_\mu\gamma^\mu - m)\psi - Z_m m\bar{\psi}\psi - Z_1 e\bar{\psi}\gamma^\mu\psi A_\mu. \tag{2.34}$$

Note that in the derivation of the path integral earlier we explicitly used the unrescaled free-field Lagrangian to get (2.11). To keep it valid we need to put all Z-dependence in the $\mathcal{L}_{\text{int}} = \mathcal{L}_{\text{int}}^0 + \mathcal{L}_{\text{ct}}$

$$\mathcal{L}_{\text{QED}} = \overbrace{-\frac{1}{4}F_{\mu\nu}F^{\mu\nu} + \bar{\psi}(i\partial_\mu\gamma^\mu - m)\psi}^{\mathcal{L}_0} \overbrace{- Z_1 e\bar{\psi}\gamma^\mu\psi A_\mu}^{\mathcal{L}_{\text{int}}^0}$$
$$\underbrace{-\frac{1}{4}(Z_3 - 1)F_{\mu\nu}F^{\mu\nu} + (Z_2 - 1)\bar{\psi}(i\partial_\mu\gamma^\mu - m)\psi - Z_m m\bar{\psi}\psi}_{\mathcal{L}_{\text{ct}}} \quad (2.35)$$

The new part $\mathcal{L}_{\text{ct}}$ is called *counterterm* Lagrangian. The Feynman rules have to be extended to include additional rules for counterterms.

Out of four renormalization parameters, only Z_2 and Z_3 are fixed by conditions (2.32). If we try to calculate them we will find out that they are divergent too (containing poles in ϵ), but those divergences cancel some of the UV-divergences coming from loop integrals in the full amplitude. That brings us to the understanding that the initial parameters in the Lagrangian (*bare* parameters) are unphysical and can even be infinite. As long as physical observables are finite the divergence of bare parameters should not concern us.

We have to provide two more renormalization conditions to fix the remaining parameters Z_m and Z_1. For instance we can demand m to be the actual mass of the fermion (the pole in the full propagator) and e be equal to the electron electric charge observed in the Thomson scattering experiment. This will give us the following conditions

$$\Sigma(\not{p} = m) = 0, \qquad -ie\Gamma^\mu(q = 0) = -ie\gamma^\mu \quad (2.36)$$

where $\Sigma(p)$ is the full one particle irreducible insertion into the fermion propagator and $\Gamma^\mu(q)$ is the full fermion-photon vertex. With (2.32) and (2.36) one can calculate perturbative expansion for renormalization constants order by order using Feynman diagrams.

The arbitrariness in the choice of renormalization conditions leads to the possibility of using different renormalization schemes. The most commonly used ones are *on-shell* (2.36) and *MS-bar*. The physical predictions of the theory are independent of the choice of renormalization scheme.

The QED as well as the rest of the Standard Model belongs to the class of renormalizable theories. One can rigorously prove that in such theories a finite number of renormalization constants is sufficient to remove all UV-divergences from all experimental predictions [60].

For completeness we list below Feynman rules for counterterm vertices

- $i(Z_2 - 1)(\slashed{q} - m) - iZ_m m$ fermion counterterm
- $-i(Z_3 - 1)(g^{\mu\nu}q^2 - q^\mu q^\nu)$ photon counterterm
- $-i(Z_1 - 1)eQ_e \gamma^\mu$ charge counterterm

2.5. Cancellation of IR divergences

After performing renormalization we are left with an amplitude which still has infrared divergences. The existence of these divergences roots in the definition of asymptotic states. It turns out that the idea of asymptotic states as isolated single particle states is not a valid abstraction for theories with massless particles (e.g. photons).

To correct this abstraction we should ask ourselves what we can and what we cannot see in the experiment. Any real detector has a finite resolution. So in principle a scattering process can be accompanied by a number of very soft (low-energy) massless particles that escaped the detection. Likewise one cannot distinguish a situation of several massless collinear particles moving together from a single massless particle with the same energy.

We have to account for this undetected particles by including additional terms in our scattering cross section definition. Which can be schematically shown as

$$d\sigma_n = \frac{1}{4|\mathbf{p}_1|\sqrt{s}} \left(|M_n|^2 \, d\Phi_n + \int_0^\delta |M_{n+1}|^2 \, d\Phi_{n+1} + \iint_0^\delta |M_{n+2}|^2 \, d\Phi_{n+2} \cdots \right)$$
(2.37)

where the integration goes over the phase-space of unobserved particles from zero up to detector resolution δ.

Let us have a closer look at these new terms. The matrix element of n particle scattering with additional photon emission can be written as

$$\mathbf{M}_n \text{ (diagram)} = M_n \times (-ie)\gamma^\mu \frac{i(\not{p} + \not{k} + m)}{(p+k)^2 - m^2}$$
(2.38)

We can see that in the limit $k \to 0$ the Eq. (2.38) can diverge. Therefore phase-space integrations in (2.37) are potentially divergent too and have to be regulated. It turns out that infrared singularities coming from virtual loop corrections and singularities from soft/collinear real emission exactly cancel each other leaving us with a finite physical cross section [43, 123, 130, 177].

This concludes our short overview of theoretical concepts one should familiarize himself with before proceeding on to the following chapters. More information can be found in one of many textbooks on quantum field theory like for instance [53, 147, 157].

3. One-loop tensor reduction

The stable numerical evaluation of tensor integrals is one of the central ingredients of any one loop Feynman diagram calculation.

The classic Passarino-Veltman reduction scheme allows to express tensor integrals in terms of basis of 4-dimensional scalar 1-, 2-, 3- and 4-point integrals with kinematic coefficients [52, 146, 162]. While it works well for processes with up to 4 external states, for 5 and more legs the numerical stability of reduction coefficients is spoiled by the appearance of inverse Gram determinants. A number of methods have been proposed to avoid Gram determinant and improve the numerical stability [37, 38, 54, 69, 75, 103, 158, 159, 163, 169, 171].

In this chapter we introduce the reader to modern reduction techniques with focus on the dimensional recurrence approach. After giving definitions for tensor integrals and kinematic determinants, we outline the straightforward Passarino-Veltman reduction scheme and explicitly show the appearance of inverse Gram determinants. Then we describe the alternative approach based on using integrals in the shifted dimension. Starting with the reduction of 5-point functions, we proceed to methods of evaluation of 4-point functions and conclude by the discussion of the reduction of triangles and bubbles.

3.1. Definitions and notation

3.1.1. Dimensionally regulated tensor integrals

We define dimensionally regulated n-point 1-loop tensor integral of rank R as:

$$I_n^{\mu_1 \ldots \mu_R} = (2\pi\mu)^{2\epsilon} \int \frac{d^d k}{i\pi^{d/2}} \frac{k^{\mu_1} \ldots k^{\mu_R}}{((k-q_1)^2 - m_1^2 + i\epsilon) \cdots ((k-q_n)^2 - m_n^2 + i\epsilon)} \quad (3.1)$$

where chords q_i are defined as

$$q_1 = p_1, \quad q_2 = p_1 + p_2, \quad q_3 = p_1 + p_2 + p_3, \quad \ldots, \quad q_n = \sum_{i=1}^{n} p_i \qquad (3.2)$$

Here $d = 4 - 2\epsilon$ is the space-time dimension and ϵ is the dimensional regulator. The Eq. 3.1 can contain single and double poles at $\epsilon = 0$ and should be thought as a Laurent series expansion near that point. The physically relevant information is contained in the finite term of this expansion, while $1/\epsilon$ and $1/\epsilon^2$ provide additional cross-checks as they have to cancel against similar terms from real emission and renormalization.

The tensor integral (3.1) is invariant under loop momenta shifts $k'_\mu = k_\mu + \xi_\mu$. We can use this freedom to put one denominator into a simple form by eliminating its chord $q_n = 0$.

In this chapter we pick an all-outgoing momenta labeling, which is shown in Fig. 3.1, this also fixes the sign of q_i's in the denominators.

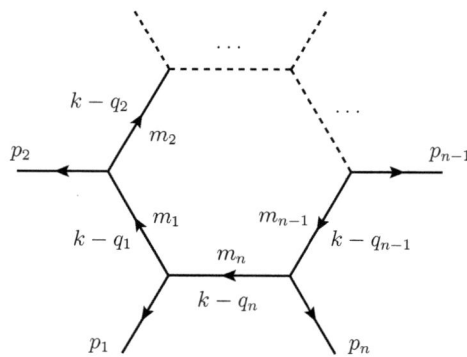

Figure 3.1.: All-outgoing momenta labeling

To specify n-point kinematics in a Lorentz-invariant fashion we use generalized Mandelstam variables s_{ij}:

$$s_{ij} = (p_i + p_j)^2, \qquad i, j = 1, \ldots, n \qquad (3.3)$$

3.1.2. Kinematic determinants

We use the Melrose [139] notation for kinematic matrices and determinants. The modified Cayley matrix of n-point kinematics:

$$Y_{ij} = m_i^2 + m_j^2 - (q_i - q_j)^2, \quad Y_{0i} = Y_{i0} = 1, \quad Y_{00} = 0, \quad i,j = 1,\ldots,n \tag{3.4}$$

and its determinant

$$()_n = \begin{vmatrix} 0 & 1 & 1 & \cdots & 1 \\ 1 & Y_{11} & Y_{12} & \cdots & Y_{1n} \\ \vdots & \vdots & \vdots & \ddots & \vdots \\ 1 & Y_{1n} & Y_{2n} & \cdots & Y_{nn} \end{vmatrix}$$

The order r cofactors of the Cayley matrix (also called signed minors)

$$\begin{pmatrix} i_1 i_2 \ldots i_r \\ j_1 j_2 \ldots j_r \end{pmatrix}_n = \det\left(Y_{\{j_1 j_2 \ldots j_r\}}^{\{i_1 i_2 \ldots i_r\}}\right) \operatorname{sgn}(i_1 i_2 \ldots i_r) \operatorname{sgn}(j_1 j_2 \ldots j_r) \prod_{k=1}^{r} (-1)^{i_k + j_k} \tag{3.5}$$

where $Y_{\{j_1 j_2 \ldots j_r\}}^{\{i_1 i_2 \ldots i_r\}}$ is an $(n+1-r) \times (n+1-r)$ matrix obtained from the Cayley matrix by discarding rows $i_1 i_2 \ldots i_r$ and columns $j_1 j_2 \ldots j_r$. The signature of permutation of indices $(i_1 i_2 \ldots i_r)$ to the ordered set is denoted as $\operatorname{sgn}(i_1 i_2 \ldots i_r)$.

The signed minors obey a multitude of algebraic relations [139] which are extensively used throughout this chapter. The most often used master identity is

$$\begin{pmatrix} \alpha \\ 0 \end{pmatrix}_n \begin{pmatrix} \alpha\,\beta \\ i\,j \end{pmatrix}_n + \begin{pmatrix} \alpha \\ i \end{pmatrix}_n \begin{pmatrix} \alpha\,\beta \\ j\,0 \end{pmatrix}_n + \begin{pmatrix} \alpha \\ j \end{pmatrix}_n \begin{pmatrix} \alpha\,\beta \\ 0\,i \end{pmatrix}_n = 0 \tag{3.6}$$

where $\alpha, \beta = 0, \ldots, n$ and $i, j = 1, \ldots, n$

See App. A for more information about signed minors.

We define the $(n-1)\times(n-1)$ Gram matrix and note the connection between Gram determinant and Cayley determinant at $q_n = 0$:

$$G_{ij}^{(n)} = 2q_i q_j, \qquad G^{(n)} \equiv \det G_{ij}^{(n)} = -()_n\big|_{q_n=0} \qquad (3.7)$$

3.2. Passarino-Veltman reduction

The first systematic approach to the evaluation of one-loop tensor integrals was developed in the work of Passarino and Veltman [146]. Their method defines tensor integrals in terms of generic scalar form-factors by separating the Lorentz structure into products of external momenta and/or metric tensors.

Any one loop tensor integral can be written in the following form:

$$I_n^{\mu_1\ldots\mu_R} = \sum_{i_1,\ldots,i_R}^{n} q_{i_1}^{[\mu_1}\cdots q_{i_R}^{\mu_R]} F_{i_1\ldots i_R}^{(n)} + \sum_{i_3,\ldots,i_R}^{n} g^{[\mu_1\mu_2} q_{i_3}^{\mu_3}\cdots q_{i_R}^{\mu_R]} F_{00 i_3\ldots i_R}^{(n)}$$
$$+ \sum_{i_5,\ldots,i_R}^{n} g^{[\mu_1\mu_2} g^{\mu_3\mu_4} q_{i_5}^{\mu_5}\cdots q_{i_R}^{\mu_R]} F_{0000 i_5\ldots i_R}^{(n)} + \cdots \qquad (3.8)$$

The square brackets here denote non-equivalent symmetrization, which gives the set of all non-equivalent permutations. It can be obtained by normalizing the full symmetrization by symmetry factors of the term. Thus we have:

$$q_{i_1}^{[\mu_1}\cdots q_{i_R}^{\mu_R]} = q_{i_1}^{\mu_1}\cdots q_{i_R}^{\mu_R} \qquad (3.9a)$$

$$g^{[\mu\nu} q_i^{\lambda]} = g^{\mu\nu} q_i^{\lambda} + g^{\mu\lambda} q_i^{\nu} + g^{\nu\lambda} q_i^{\mu} \qquad (3.9b)$$

$$g^{[\mu\nu} q_i^{\lambda} q_j^{\sigma]} = g^{\mu\nu} q_i^{\lambda} q_j^{\sigma} + g^{\mu\lambda} q_i^{\nu} q_j^{\sigma} + g^{\mu\sigma} q_i^{\nu} q_j^{\lambda} + g^{\nu\lambda} q_i^{\mu} q_j^{\sigma} + g^{\nu\sigma} q_i^{\mu} q_j^{\lambda} + g^{\lambda\sigma} q_i^{\mu} q_j^{\nu} \qquad (3.9c)$$

$$g^{[\mu\nu} g^{\lambda\sigma]} = g^{\mu\nu} g^{\lambda\sigma} + g^{\mu\lambda} g^{\nu\sigma} + g^{\mu\sigma} g^{\nu\lambda} \qquad (3.9d)$$

$$g^{[\mu\nu} q_i^{\lambda} q_j^{\sigma} q_k^{\rho]} = g^{\mu\nu} q_i^{\lambda} q_j^{\sigma} q_k^{\rho} + g^{\mu\lambda} q_i^{\nu} q_j^{\sigma} q_k^{\rho} + g^{\mu\sigma} q_i^{\nu} q_j^{\lambda} q_k^{\rho} + g^{\mu\rho} q_i^{\nu} q_j^{\lambda} q_k^{\sigma} + g^{\nu\lambda} q_i^{\mu} q_j^{\sigma} q_k^{\rho}$$
$$+ g^{\nu\sigma} q_i^{\mu} q_j^{\lambda} q_k^{\rho} + g^{\nu\rho} q_i^{\mu} q_j^{\lambda} q_k^{\sigma} + g^{\lambda\sigma} q_i^{\mu} q_j^{\nu} q_k^{\rho} + g^{\lambda\rho} q_i^{\mu} q_j^{\nu} q_k^{\sigma} + g^{\sigma\rho} q_i^{\mu} q_j^{\nu} q_k^{\lambda} \qquad (3.9e)$$

$$g^{[\mu\nu} g^{\lambda\sigma} q_i^{\rho]} = g^{\mu\nu} g^{\lambda\sigma} q_i^{\rho} + g^{\mu\nu} g^{\lambda\rho} q_i^{\sigma} + g^{\mu\nu} g^{\rho\sigma} q_i^{\lambda} + g^{\mu\rho} g^{\lambda\sigma} q_i^{\nu} + g^{\rho\nu} g^{\lambda\sigma} q_i^{\mu} \qquad (3.9f)$$

The scalar functions $F^{(n)}_{i_1...i_R}$ are tensor form-factors. Tensor integrals and their form-factors are often denoted by capital letters [146]

$$F^{(1)}_{...} = A_{...}, \quad F^{(2)}_{...} = B_{...}, \quad F^{(3)}_{...} = C_{...}, \quad F^{(4)}_{...} = D_{...}, \quad F^{(5)}_{...} = E_{...} \quad (3.10)$$

For convenience we list the explicit form of Eq. (3.8) for tensors up to rank 5:

$$I^\mu_n = \sum_{i=1}^n q_i^\mu F_i^{(n)} \quad (3.11a)$$

$$I^{\mu\nu}_n = \sum_{i,j=1}^n q_i^\mu q_j^\nu F_{ij}^{(n)} + g^{\mu\nu} F_{00}^{(n)} \quad (3.11b)$$

$$I^{\mu\nu\lambda}_n = \sum_{i,j,k=1}^n q_i^\mu q_j^\nu q_k^\lambda F_{ijk}^{(n)} + \sum_{i=k}^n g^{[\mu\nu} q_k^{\lambda]} F_{00k}^{(n)} \quad (3.11c)$$

$$I^{\mu\nu\lambda\sigma}_n = \sum_{i,j,k,l=1}^n q_i^\mu q_j^\nu q_k^\lambda q_l^\sigma F_{ijkl}^{(n)} + \sum_{k,l=1}^n g^{[\mu\nu} q_k^\lambda q_l^{\sigma]} F_{00kl}^{(n)} + g^{[\mu\nu} g^{\lambda\sigma]} F_{0000}^{(n)} \quad (3.11d)$$

$$I^{\mu\nu\lambda\sigma\rho}_n = \sum_{i,j,k,l,m=1}^n q_i^\mu q_j^\nu q_k^\lambda q_l^\sigma q_m^\rho F_{ijklm}^{(n)}$$
$$+ \sum_{k,l,m=1}^n g^{[\mu\nu} q_k^\lambda q_l^\sigma q_m^{\rho]} F_{00klm}^{(n)} + \sum_{m=1}^n g^{[\mu\nu} g^{\lambda\sigma} q_m^{\rho]} F_{0000m}^{(n)} \quad (3.11e)$$

One may notice that after choosing $q_n = 0$ it becomes possible to simplify the contraction of tensor integral (3.1) with $q_{s,\mu}$ using this relation:

$$2q_{s,\mu} k^\mu = (q_s^2 + m_n^2 - m_s^2) - ((k - q_s)^2 - m_s^2) + (k^2 - m_n^2) \quad (3.12)$$

The first term does not contain the loop momentum and corresponds to to a tensor integral with reduced rank. The last two terms cancel the sth and n^{th} denominators and result in integrals with reduced rank and number of legs.

By introducing the notation $I^{\mu_1...\mu_R,s}_n$ for the integral with canceled s^{th} propagator, we can write a contraction of the rank-R tensor integral (3.1) with $2q_{s,\mu}$ and $g_{\mu\nu}$ as:

$$2q_{s,\mu_1}I_n^{\mu_1\mu_2\cdots\mu_R} = (q_s^2 + m_n^2 - m_s^2)I_n^{\mu_2\cdots\mu_R} - I_{n-1}^{\mu_2\cdots\mu_R,s} + I_{n-1}^{\mu_2\cdots\mu_R,n} \qquad (3.13a)$$

$$g_{\mu_1\mu_2}I_n^{\mu_1\mu_2\cdots\mu_R} = m_n^2 I_n^{\mu_3\cdots\mu_R} + I_{n-1}^{\mu_3\cdots\mu_R,n} \qquad (3.13b)$$

Substituting form-factor expansions (3.8) into (3.13a) we get a system of equations

$$\sum_{i_2,\ldots,i_R=1}^{n-1} q_{i_2}^{\mu_2} \cdots q_{i_R}^{\mu_R} \times \left(\sum_{k=1}^{n-1} G_{sk}^{(n)} F_{ki_2\ldots i_R}^{(n)} + 2\sum_{r=2}^{R} \delta_{si_r} F_{00i_2\ldots i_{r-1}i_{r+1}\ldots i_R}^{(n)} \right) + \cdots$$

$$= \sum_{i_2,\ldots,i_R=1}^{n-1} q_{i_2}^{\mu_2} \cdots q_{i_R}^{\mu_R} \times \left((q_s^2 + m_n^2 - m_s^2) F_{i_2\ldots i_R}^{(n)} - F_{i_2\ldots i_R}^{(n-1),s} + F_{i_2\ldots i_R}^{(n-1),n} \right) + \cdots$$

$$(3.14a)$$

and the same for (3.13b)

$$\sum_{i_3,\ldots,i_R=1}^{n-1} q_{i_3}^{\mu_3} \cdots q_{i_R}^{\mu_R} \times \left(\sum_{k,m=1}^{n-1} \frac{1}{2} G_{km}^{(n)} F_{kmi_3\ldots i_R}^{(n)} + (D+R-2+R') F_{00i_3\ldots i_R}^{(n)} \right) + \cdots$$

$$= \sum_{i_3,\ldots,i_R=1}^{n-1} q_{i_3}^{\mu_3} \cdots q_{i_R}^{\mu_R} \times \left(m_n^2 F_{i_3\ldots i_R}^{(n)} + F_{i_3\ldots i_R}^{(n-1),n} \right) + \cdots \qquad (3.14b)$$

where $G_{ij}^{(n)}$ is the Gram matrix (3.7) and R' is the number of non-zero indices among $i_3, \ldots, i_R$.

First we multiply (3.14a) by the inverse Gram matrix $\left(G^{(n)}\right)^{-1}$ and solve it for $F_{i_1\ldots i_R}^{(n)}$

$$F_{i_1\ldots i_R}^{(n)} = \sum_{s=1}^{n-1} \left(G^{(n)}\right)^{-1}_{i_1 s} \Bigg(-2\sum_{r=2}^{R} \delta_{si_r} F_{00i_2\ldots i_{r-1}i_{r+1}\ldots i_R}^{(n)}$$

$$+ (q_s^2 + m_n^2 - m_s^2) F_{i_2\ldots i_R}^{(n)} - F_{i_2\ldots i_R}^{(n-1),s} + F_{i_2\ldots i_R}^{(n-1),n} \Bigg) \quad (3.15)$$

Now we can substitute (3.15) to (3.14b) and get a relation for $F^{(n)}_{00i_3...i_R}$:

$$F^{(n)}_{00i_3...i_R} = \frac{1}{2(D+R-1-n)}\left[2m_n^2 F^{(n)}_{i_3...i_R} + 2F^{(n-1),n}_{i_3...i_R} \right.$$
$$\left. - \sum_{s=1}^{n-1}\left((q_s^2 + m_n^2 - m_s^2)F^{(n)}_{si_3...i_R} - F^{(n-1),s}_{si_3...i_R} + F^{(n-1),n}_{si_3...i_R}\right)\right] \quad (3.16)$$

As we can see (3.15) and (3.16) express rank-R n-point tensor form-factors in terms of lower n and/or lower rank form-factors. Therefore one can apply them recursively to write any tensor integral in terms of scalar integrals, which are known analytically [87].

In this derivation we assumed that the Gram matrix is non-singular, which is generally true for $n \leq 5$ with non-special kinematics. The inverse can be written as:

$$\left(\mathbf{G}^{(n)}\right)^{-1}_{ij} = \frac{\mathrm{adj}\left(\mathbf{G}^{(n)}\right)_{ij}}{\det \mathbf{G}^{(n)}} \quad (3.17)$$

where the numerator is the adjugate or cofactor matrix and the denominator is the Gram determinant, which we will further denote as $G^{(n)}$.

Each step of recurrence (3.15) introduces a single power of the Gram determinant in the denominator. This can potentially affect the numerical accuracy when the Gram determinant comes close to zero. For up to $n = 4$ this happens only near various thresholds on the edge of the physical phase-space.

For instance for massless kinematics $G^{(4)} = -s_{12}s_{23}(s_{12}+s_{23})$ becomes small in soft and collinear regions which are normally excluded by experimental cuts.

Unfortunately, already for $n = 5$, the regions where $G^{(5)}$ and $G^{(4)}$ can become small or zero overlap with the physical phase space. Even in the simplest massless case

$$G^{(5)} = -s_{12}^2(s_{15} - s_{23})^2 - \left(s_{23}s_{34} + (s_{15} - s_{34})s_{45}\right)^2 +$$
$$+ 2s_{12}\left(s_{23}s_{34}(s_{23} - s_{45}) + s_{15}^2 s_{45} - s_{15}(s_{34}s_{45} + s_{23}(s_{34} + s_{45}))\right) \quad (3.18)$$

The situation gets worse with adding masses and/or increasing number of legs.

For $n \geq 6$ the Gram determinant is identical to zero $G^{(n)} \equiv 0|_{n \geq 6}$. Due to the dimensionality of space time we can't have more than four linearly independent momenta q_i.

It can be shown that any rank-R $n \geq 6$ tenor integral can be reduced iteratively to tensor 5-point integrals without introducing inverse Gram determinants [30, 38, 75, 82, 96]. Therefore the tensor pentagon form-factors are the most complicated objects which we need to know how to evaluate for small Gram determinants.

For instance, using the signed minors notation defined in (3.5):

$$I_6^{\mu_1 \mu_2 \ldots \mu_R} = - \sum_{i=1}^{6} q_i^{\mu_1} \sum_{s=1}^{6} \frac{\binom{0s}{0i}_6}{\binom{0}{0}_6} I_5^{\mu_2 \ldots \mu_R, s} \tag{3.19}$$

3.3. Dimensional recurrence method

One possible solution to the Gram determinant problem is using different set of basis integrals. It turns out that coefficients of one-loop tensor integrals can be directly related to scalar integrals in different space-time dimensions [69]. And that integrals satisfy certain recursion relations [163]. This ideas have been further developed by several groups [27, 30, 37, 38, 82, 94, 96, 103].

The basis of higher-dimensional integrals is free of Gram determinants, but numerically stable analytic expressions are hard to obtain and are known only for the massless case. Therefore several semi-numerical approaches have been proposed [41, 88, 102].

In this section we will focus on the findings of [81, 82, 83, 94] and outline the most important steps of the reduction scheme.

3.3.1. Scalar integrals in shifted dimension

We start by writing down the Davydychev formula [69] explicitly for form-factors of tensor integrals up to rank 5

$$F_i^{(n)} = -I_{n,i}^{[2]} \tag{3.20a}$$

$$F_{ij}^{(n)} = n_{ij} I_{n,ij}^{[4]} \qquad F_{00}^{(n)} = -\frac{1}{2} I_n^{[2]} \tag{3.20b}$$

$$F_{ijk}^{(n)} = -n_{ijk} I_{n,ijk}^{[6]} \qquad F_{00k}^{(n)} = \frac{1}{2} I_{n,k}^{[4]} \tag{3.20c}$$

$$F_{ijkl}^{(n)} = n_{ijkl} I_{n,ijkl}^{[8]} \qquad F_{00kl}^{(n)} = -\frac{1}{2} n_{kl} I_{n,kl}^{[6]} \qquad F_{0000}^{(n)} = \frac{1}{4} I_n^{[4]} \tag{3.20d}$$

$$F_{ijklm}^{(n)} = -n_{ijklm} I_{n,ijklm}^{[10]} \qquad F_{00klm}^{(n)} = \frac{1}{2} n_{km} I_{n,klm}^{[8]} \qquad F_{0000m}^{(n)} = -\frac{1}{4} I_{n,m}^{[6]} \tag{3.20e}$$

where $I_{n,i_1 i_2 \ldots}^{[2l], s_1 s_2 \ldots}$ is a generalized scalar loop integral in shifted dimension with shifted powers of denominators:

$$I_{n,i_1 i_2 \ldots}^{[2l], s_1 s_2 \ldots} = (2\pi\mu)^{2\epsilon} \int \frac{d^{d+2l} k}{i\pi^{d/2+l}} \prod_{r=1}^n \frac{1}{((k-q_r)^2 - m_r^2 + i\epsilon)^{1+\delta_{r i_1} + \delta_{r i_2} + \cdots - \delta_{r s_1} - \delta_{r s_2} - \cdots}} \tag{3.21}$$

The symbols $n_{i_1 i_2 \ldots}$ in (3.20) are shorthand notations for combinatorial factors introduced in [82]. They are equal to the products of factorials of numbers of equal indices: $n_{ijk} = 1!1!1!$, $n_{iik} = 1!2!$, $n_{iii} = 3!$. They are related to another set of indexed objects $v_{i_1 i_2 \ldots}$:

$$n_{ij} = v_{ij} \qquad v_{ij} = 1 + \delta_{ij} \tag{3.22a}$$

$$n_{ijk} = v_{ij} v_{ijk} \qquad v_{ijk} = 1 + \delta_{ik} + \delta_{jk} \tag{3.22b}$$

$$n_{ijkl} = v_{ij} v_{ijk} v_{ijkl} \qquad v_{ijkl} = 1 + \delta_{il} + \delta_{jl} + \delta_{kl} \tag{3.22c}$$

$$n_{ijklm} = v_{ij} v_{ijk} v_{ijkl} v_{ijklm} \qquad v_{ijklm} = 1 + \delta_{im} + \delta_{jm} + \delta_{km} + \delta_{lm} \tag{3.22d}$$

where δ_{ij} is Kronecker delta symbol.

These combinatorial factors appear naturally through the application of recurrence relations between generic scalar loop integrals [96, 163]. Introducing

ν_j for the power of the jth denominator (index), we get for the dimension shifting recurrence

$$()_n(d + 2l - \sum_{i=1}^{n}\nu_i + 1)I_n^{[2(l+1)]} = \binom{0}{0}_n I_n^{[2l]} - \sum_{s=1}^{n}\binom{0}{s}_n \mathbf{s}^{-}I_n^{[2l]} \qquad (3.23)$$

and for the combined recurrence in dimension and index

$$()_n \nu_j \mathbf{j}^{+} I_n^{[2(l+1)]} = -\binom{j}{0}_n I_n^{[2l]} + \sum_{s=1}^{n}\binom{j}{s}_n \mathbf{s}^{-}I_n^{[2l]} \qquad (3.24)$$

where $\mathbf{s}^{-}$ is s^{th} denominator power lowering operator and $\mathbf{j}^{+}$ is j^{th} denominator power raising operator.

3.3.2. Reduction of 5-point functions

The reduction of a scalar pentagon to scalar boxes has been known for many years [139, 169] and follows directly from (3.23)

$$()_n(d-4)I_5^{[2]} = \binom{0}{0}_5 I_5 - \sum_{s=1}^{5}\binom{0}{s}_5 I_4^s \qquad (3.25)$$

Since $I_5^{[2]}$ is finite its product with $d - 4 = -2\epsilon$ starts at $O(\epsilon)$ and we get for the finite part of the scalar pentagon:

$$E_0 \equiv I_5 = \frac{1}{\binom{0}{0}_5} \sum_{s=1}^{5}\binom{0}{s}_5 I_4^s \qquad (3.26)$$

The vector form-factor (3.20a) can be similarly obtained from (3.24)

$$-()_n F_i^{(5)} = -\binom{i}{0}_5 I_5 + \sum_{s=1}^{5}\binom{i}{s}_5 I_4^s \qquad (3.27)$$

substituting (3.26)

$$E_i \equiv F_i^{(5)} = \left[\frac{\binom{i}{0}_5\binom{0}{s}_5}{\binom{0}{0}_5\binom{0}{0}_5} + \frac{\binom{i}{s}_5}{\binom{0}{0}_5}\right] \sum_{s=1}^{5} I_4^s \qquad (3.28)$$

This expression for the vector pentagon form-factor contains the Gram determinant $G^{(5)} = ()_5$ in the denominator. It can be eliminated by using the signed minors' identity (A.6)

$$\binom{0}{0}_5 \binom{s}{i}_5 = \binom{0s}{0i}_5 ()_5 + \binom{0}{i}_5 \binom{s}{0}_5 \qquad (3.29)$$

$$E_i = -\frac{1}{\binom{0}{0}_5} \sum_{s=1}^{5} \binom{0s}{0i}_5 I_4^s \qquad (3.30)$$

One observes the appearance of the Cayley determinant $\binom{0}{0}_5$ in the denominator of the reduction formulae (3.26) and (3.30). This raises the question whether it can become small and affect the numerical stability. Generally it is preferable to have the Cayley determinant in the denominator instead of the regular Gram determinant. For instance in the massless case, unlike the Gram determinant (3.18), the Cayley determinant never becomes zero within the physical phase-space:

$$\binom{0}{0}_5 = -2s_{12}s_{15}s_{23}s_{34}s_{45}, \qquad \binom{0}{0}_4 = s^2 t^2 \qquad (3.31)$$

In case of large Gram determinants one can always use a Passarino-Veltman-like reduction. Which has the advantage of reducing the number of form-factors by eliminating $g^{\mu\nu}$. This is possible due to the fact that in case of 5-point kinematics q_i's form a complete basis in the 4-dimensional space-time

$$g^{\mu\nu} = 2 \sum_{i,j=1}^{5} \frac{\binom{i}{j}_5}{()_5} q_i^\mu q_j^\mu \qquad (3.32)$$

While it is still possible to use (3.23) and (3.24) directly for the higher rank tensors, the fact that they introduce an inverse Gram determinant on every application calls for a better approach. It turns out that it is more convenient to start with explicitly solved recurrence relations [83] similar to (3.19)

$$I_5^{\mu_1\mu_2\cdots\mu_R} = \sum_{i=1}^{5} q_i^{\mu_1} \left[\frac{\binom{0}{i}_5}{()_5} I_5^{\mu_2\cdots\mu_R} - \sum_{s=1}^{5} \frac{\binom{s}{i}_5}{()_5} I_4^{\mu_2\cdots\mu_R,s} \right] \qquad (3.33)$$

after using (3.29)

$$I_5^{\mu_1\mu_2\cdots\mu_R} = \sum_{i=1}^{5} q_i^{\mu_1} \left[\frac{\binom{0}{i}_5}{()_5} \overbrace{\left(I_5^{\mu_2\cdots\mu_R} - \sum_{s=1}^{5} \frac{\binom{s}{0}_5}{\binom{0}{0}_5} I_4^{\mu_2\cdots\mu_R,s} \right)}^{T_5^{\mu_2\cdots\mu_R}} - \sum_{s=1}^{5} \frac{\binom{0s}{0i}_5}{\binom{0}{0}_5} I_4^{\mu_2\cdots\mu_R,s} \right] \qquad (3.34)$$

In this form it still contains a Gram determinant $()_5$ and we have to transform $T_5^{\mu_2\cdots\mu_R}$ to be able to apply the following signed minors' algebra relation

$$\binom{s}{i}_5 \frac{\binom{0}{j}_5}{()_5} = -\binom{0i}{sj}_5 + \binom{s}{0}_5 \frac{\binom{i}{j}_5}{()_5} \qquad (3.35)$$

where the last term will become $g^{\mu\nu}$ after use of (3.32).

Pentagon rank 2

We start with the $T_5^{\mu_2\cdots\mu_R}$ part of master formula (3.34) and feed the recursion with the explicitly obtained expression for the vector pentagon (3.30) and the standard expression (3.20a) for a vector box

$$T_5^{\mu_2} = I_5^{\mu_2} - \sum_{s=1}^{5} \frac{\binom{s}{0}_5}{\binom{0}{0}_5} I_4^{\mu_2,s} = \sum_{j=1}^{5} q_j^{\mu_2} \left[E_j - \sum_{s=1}^{5} \frac{\binom{s}{0}_5}{\binom{0}{0}_5} D_j^s \right]$$

$$= \sum_{j=1}^{5} q_j^{\mu_2} \sum_{s=1}^{5} \frac{1}{\binom{0}{0}_5} \left[-\binom{0s}{0j}_5 I_4^s + \binom{s}{0}_5 I_{4,j}^{[2],s} \right]$$

applying (3.24) to the last term

$$= \sum_{j=1}^{5} q_j^{\mu_2} \sum_{s=1}^{5} \frac{1}{\binom{0}{0}_5} \left[-\binom{0s}{0j}_5 I_4^s + \binom{s}{0}_\varepsilon \left\{ -\frac{\binom{js}{0s}_5}{\binom{s}{s}_5} I_4^s + \sum_{t=1}^{5} \frac{\binom{js}{ts}_5}{\binom{s}{s}_5} I_3^{st} \right\} \right] \quad (3.36)$$

Here we should note Gram determinants of 4-point subkinematics $\binom{s}{s}_5$ which came from vector boxes reduction. With help of two special cases of (3.6) the first term corresponding to E_i can be canceled in (3.36)

$$\binom{s}{0}_5 \binom{0\,s}{i\,s}_5 = \binom{s}{i}_5 \binom{0\,s}{0\,s}_5 - \binom{s}{s}_5 \binom{0\,s}{0\,i}_5 \quad (3.37)$$

$$\binom{s}{0}_5 \binom{t\,s}{i\,s}_5 = \binom{s}{i}_5 \binom{t\,s}{0\,s}_5 - \binom{s}{s}_5 \binom{t\,s}{0\,i}_5 \quad (3.38)$$

$$T_5^{\mu_2} = -\sum_{j=1}^{5} q_j^{\mu_2} \sum_{s=1}^{5} \frac{1}{\binom{0}{0}_5} \binom{s}{j}_5 \overbrace{\left[-\frac{\binom{0s}{0s}_5}{\binom{s}{s}_5} I_4^s + \sum_{t=1}^{5} \frac{\binom{0s}{ts}_5}{\binom{s}{s}_5} I_3^{st} \right]}^{I_4^{[2],s}+O(\epsilon)} - \sum_{j=1}^{5} q_j^{\mu_2} \sum_{s,t=1}^{5} \overbrace{\binom{ts}{0j}}^{0} I_3^{st} \quad (3.39)$$

Note how scalar box and scalar triangles are combined into the finite six-dimensional box $I_4^{[2],s}$ by reverse application of (3.23). The last term is summed to zero due to the antisymmetry property of signed minors (A.4).

Now we are ready to plug $T_5^{\mu_2}$ into (3.34) and use (3.35) to get rid of the Gram determinant $()_5$

$$I_5^{\mu_1\mu_2} = \sum_{i=1}^{5} q_i^{\mu_1} \left[-\frac{\binom{0}{i}_5}{\binom{0}{0}_5} \sum_{j=1}^{5} q_j^{\mu_2} \sum_{s=1}^{5} \frac{\binom{s}{j}_5}{\binom{0}{0}_5} I_4^{[2],s} - \sum_{s=1}^{5} \frac{\binom{0s}{0i}_5}{\binom{0}{0}_5} I_4^{\mu_2,s} \right]$$

$$= \sum_{i,j=1}^{5} q_i^{\mu_1} q_j^{\mu_2} \sum_{s=1}^{5} \left[-\frac{\binom{0}{i}_5 \binom{s}{j}_5}{\binom{0}{0}_5 \binom{0}{0}_5} I_4^{[2],s} + \frac{\binom{0s}{0i}_5}{\binom{0}{0}_5} I_{4,j}^{[2],s} \right]$$

$$= \sum_{i,j=1}^{5} q_i^{\mu_1} q_j^{\mu_2} \sum_{s=1}^{5} \left[\left\{ \frac{\binom{0j}{si}_5}{\binom{0}{0}_5} - \frac{\binom{s}{0}_5 \binom{i}{j}_5}{\binom{0}{0}_5 \binom{s}{s}_5} \right\} I_4^{[2],s} + \frac{\binom{0s}{0i}_5}{\binom{0}{0}_5} I_{4,j}^{[2],s} \right] \quad (3.40)$$

Summing the second term with the help of (3.32) we get the form-factors of the rank 2 pentagon

$$E_{00} = \sum_{s=1}^{5} -\frac{1}{2}\frac{1}{\binom{0}{0}_5}\binom{s}{0}_5 I_4^{[2],s} \tag{3.41}$$

$$E_{ij} = \sum_{s=1}^{5} \frac{1}{\binom{0}{0}_5}\left[\binom{0j}{si}_5 I_4^{[2],s} + \binom{0s}{0i}_5 I_{4,j}^{[2],s}\right] \tag{3.42}$$

Pentagon rank 3

Following the algorithm we start with (3.34). Unlike the vector case now we have a term proportional to $g^{\mu\nu}$, in which E_{00} cancels according to (3.41) and (3.20b)

$$\begin{aligned}
T_5^{\mu_2\mu_3} &= I_5^{\mu_2\mu_3} - \sum_{s=1}^{5} \frac{\binom{s}{0}_5}{\binom{0}{0}_5} I_4^{\mu_2\mu_3,s} \\
&= g^{\mu_2\mu_3}\left[E_{00} - \sum_{s=1}^{5}\frac{\binom{s}{0}_5}{\binom{0}{0}_5}D_{00}^s\right] + \sum_{j,k=1}^{5} q_j^{\mu_2}q_k^{\mu_3}\left[E_{jk} - \sum_{s=1}^{5}\frac{\binom{s}{0}_5}{\binom{0}{0}_5}D_{jk}^s\right] \\
&= \sum_{j,k=1}^{5} q_j^{\mu_2}q_k^{\mu_3}\sum_{s=1}^{5}\frac{1}{\binom{0}{0}_5}\left[\binom{0k}{sj}_5 I_4^{[2],s} + \binom{0s}{0j}_5 I_{4,k}^{[2],s} - \binom{s}{0}_5 \nu_{jk}I_{4,jk}^{[4],s}\right]
\end{aligned} \tag{3.43}$$

Expanding the last term with recurrence (3.24) and using (3.37) and (3.38) we can see that terms coming from E_{jk} cancel.

$$T_5^{\mu_2\mu_3} = \sum_{j,k=1}^{5} q_j^{\mu_2}q_k^{\mu_3}\sum_{s=1}^{5}\frac{1}{\binom{0}{0}_5}\left[\binom{s}{j}_5\frac{\binom{0s}{0s}_5}{\binom{s}{s}_5}I_{4,k}^{[2],s} - \binom{s}{k}_5\frac{\binom{js}{0s}_5}{\binom{s}{s}_5}I_4^{[2],s} - \binom{s}{0}_5\sum_{5t=1}^{5}\frac{\binom{st}{sj}_5}{\binom{s}{s}_5}I_{3,k}^{[2],st}\right]$$

Now we have to rewrite this expression without explicit Gram determinants $\binom{s}{s}_5$. As before, first we recurse with (3.24) down to the integrals with no lower indices, then use signed minor identities to separate one index from the rest. And finally collect everything back with reverse application of (3.24) which results in integrals with one index less

$$T_5^{\mu_2\mu_3} = \sum_{j,k=1}^{5} q_j^{\mu_2} q_k^{\mu_3} \sum_{s=1}^{5} \frac{1}{\binom{0}{0}_5} \left[\binom{s}{j}_5 \frac{\binom{0s}{0s}_5}{\binom{s}{s}_5} \left\{ -\frac{\binom{ks}{0s}_5}{\binom{s}{s}_5} I_4^s + \sum_{t=1}^{5} \frac{\binom{ks}{ts}_5}{\binom{s}{s}_5} I_3^{st} \right\} \right.$$
$$\left. - \binom{s}{k}_5 \frac{\binom{js}{0s}_5}{\binom{s}{s}_5} I_4^{[2],s} - \binom{s}{0}_5 \sum_{t=1}^{5} \frac{\binom{st}{sj}_5}{\binom{s}{s}_5} \left\{ -\frac{\binom{kst}{0st}_5}{\binom{st}{st}_5} I_3^{st} + \sum_{u=1}^{5} \frac{\binom{kst}{ust}_5}{\binom{st}{st}_5} I_2^{stu} \right\} \right]$$
(3.44)

After applying (3.37), (3.38), (3.35) and their extensions, we can combine everything into the form of (3.39)

$$T_5^{\mu_2\mu_3} = \sum_{j,k=1}^{5} q_j^{\mu_2} q_k^{\mu_3} \sum_{s=1}^{5} \frac{1}{\binom{0}{0}_5} \left[\binom{s}{j} I_{4,k}^{[4],s} + \binom{s}{k} I_{4,j}^{[4],s} + O(\epsilon) \right] \quad (3.45)$$

Inserting the above to the (3.34) and using (3.35) with (3.32) we get the form-factors of the rank 3 pentagon (including symmetrization of E_{00k})

$$E_{00k} = \sum_{s=1}^{5} \frac{1}{3} \frac{1}{\binom{0}{0}_5} \left[\binom{s}{0} I_{4,k}^{[4],s} + \frac{1}{2} \binom{0s}{0k}_5 I_4^{[2],s} \right] \quad (3.46)$$

$$E_{ijk} = \sum_{s=1}^{5} \frac{1}{\binom{0}{0}_5} \left[-\binom{0j}{si}_5 I_{4,k}^{[4],s} - \binom{0k}{si}_5 I_{4,j}^{[4],s} - \binom{0s}{0i}_5 \nu_{jk} I_{4,jk}^{[4],s} \right] \quad (3.47)$$

where the last terms in both form-factors come from (3.20b) applied to the last term of (3.34).

Higher rank pentagons

One can continue the recursion outlined in the previous subsections to obtain the expressions for form-factors of pentagons of rank 4 and rank 5.

On each step the relation (3.24) together with signed minor identities will allow to cancel lower rank tensor coefficients. Then the recurrence (3.24) is used to write everything in terms of index-less scalar integrals. The coefficients of scalar integrals are transformed with the signed minor algebra in order to be combined into scalar boxes with one lower index less. Finally the $\binom{i}{j}$ terms with help of (3.35) absorb Gram determinants into $g^{\mu\nu}$.

The result for rank 4 pentagon [94]:

$$E_{0000} = \sum_{s=1}^{5} \frac{1}{4} \frac{\binom{s}{0}_5}{\binom{0}{0}_5} I_4^{[4],s} \tag{3.48}$$

$$E_{00kl} = \sum_{s=1}^{5} -\frac{1}{4} \frac{1}{\binom{0}{0}_5} \left[\binom{0k}{sl}_5 I_4^{[4],s} + \binom{0s}{0k}_5 I_{4,l}^{[4],s} + \binom{s}{0}_5 \nu_{kl} I_{4,kl}^{[6],s} \right] \tag{3.49}$$

$$E_{ijkl} = \sum_{s=1}^{5} \frac{1}{\binom{0}{0}_5} \left[\binom{0k}{sl}_5 \nu_{ij} I_{4,ij}^{[6],s} + \binom{0i}{sl}_5 \nu_{kj} I_{4,kj}^{[6],s} + \binom{0j}{sl}_5 \nu_{ik} I_{4,ik}^{[6],s} + \binom{0s}{0l}_5 n_{ijk} I_{4,ijk}^{[6],s} \right] \tag{3.50}$$

and rank 5 pentagon

$$E_{0000m} = \sum_{s=1}^{5} -\frac{1}{20} \frac{1}{\binom{0}{0}_5} \left[\binom{0s}{0m}_5 I_4^{[4],s} + d \binom{s}{0}_5 I_{4,i}^{[6],s} \right] \tag{3.51}$$

$$E_{000klm} = \sum_{s=1}^{5} \frac{1}{40} \frac{1}{\binom{0}{0}_5} \left[3d \binom{0k}{sm}_5 I_{4,l}^{[6],s} + 3d \binom{0l}{sm}_5 I_{4,k}^{[6],s} \right.$$
$$\left. + 12 \binom{0s}{0m}_5 \nu_{kl} I_{4,kl}^{[6],s} + 8 \binom{s}{0}_5 n_{klm} I_{4,klm}^{[8],s} \right] \tag{3.52}$$

$$E_{ijklm} = \sum_{s=1}^{5} -\frac{1}{\binom{0}{0}_5} \left[\binom{0l}{sm}_5 \nu_{ij} I_{4,ijk}^{[8],s} + \binom{0i}{sm}_5 \nu_{kj} I_{4,ljk}^{[8],s} \right.$$
$$\left. + \binom{0j}{sm}_5 \nu_{ik} I_{4,ilk}^{[8],s} + \binom{0k}{sm}_5 \nu_{ik} I_{4,ijl}^{[8],s} + \binom{0s}{0m}_5 n_{ijkl} I_{4,ijkl}^{[8],s} \right] \tag{3.53}$$

3.3.3. Reduction of 4-point functions

In the previous subsection we have seen how one can derive a representation for pentagon form-factors which is free of Gram determinants $()_5$. However unlike 5-point Gram determinants $()_5$, the 4-point ones are still present though hidden inside scalar box integrals with shifted dimension and powers of denominators.

The set of functions we used to express our answer is still too generic and can be further reduced.

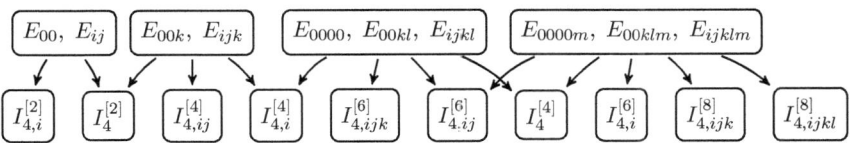

Figure 3.2.: Basis integrals for tensor pentagons reduction

The direct application of (3.24) introduces inverse Gram determinant on every step of the recursion. For the box with one index we get (see App. B for the full list)

$$I^{[2(l+1)],s}_{4,i} = \frac{1}{\binom{s}{s}_5}\left[-\binom{0s}{is}I^{[2l],s}_4 + \sum_{t=1}^{4}\binom{ts}{is}I^{[2l],st}_3\right] \qquad (3.54)$$

This recursion is very efficient because each step reduces both the dimension and the number of indices of the integral. The resulting expressions can be used if the Gram determinant $\binom{s}{s}_5$ is not small.

In case of a small 4-point Gram determinant, the fast recurrence is not suitable due to numerical instabilities. It is possible to combine basic recurrence relations with the help of signed minor identities to obtain new relations free of Gram determinants.

We start with (3.24) and apply (3.35) to the second term

$$()_n v_j \mathbf{j}^+ I^{[2(l+1)]}_n = -\binom{j}{0}_n I^{[2l]}_n + \sum_{s=1}^{n}\frac{1}{\binom{0}{0}_n}\left[\binom{0s}{0j}_n ()_n + \binom{0}{j}_n\binom{s}{0}_n\right]\mathbf{s}^- I^{[2l]}_n$$

$$v_j \mathbf{j}^+ I^{[2(l+1)]}_n = -\frac{\binom{j}{0}_n}{\binom{0}{0}_n}\left[\binom{0}{0}_n I^{[2l]}_n - \sum_{s=1}^{n}\frac{\binom{s}{0}_n}{()_n}\mathbf{s}^- I^{[2l]}_n\right] + \frac{1}{\binom{0}{0}_n}\sum_{s=1}^{n}\binom{0s}{0j}_n \mathbf{s}^- I^{[2l]}_n$$

35

using (3.23) on the bracket contents

$$\begin{pmatrix} 0 \\ 0 \end{pmatrix}_n v_j \mathbf{j}^+ I_n^{[2(l+1)]} = -\begin{pmatrix} j \\ 0 \end{pmatrix}_n (d+2l-\sum_{i=1}^n \nu_i + 1) I_n^{[2(l+1)]} + \sum_{s=1}^n \begin{pmatrix} 0s \\ 0j \end{pmatrix}_n \mathbf{s}^- I_n^{[2l]} \quad (3.55)$$

Now for the box with one index we get (see App. B for the full list)

$$I_{4,i}^{[2(l+1)]} = \frac{1}{\binom{0}{0}} \left[-(d+2l-3)\binom{0}{i} I_4^{[2(l+1)]} + \sum_{t=1}^4 \binom{0t}{0i} I_3^{[2l],t} \right] \quad (3.56)$$

The relation (3.55) contains no Gram determinant, and as mentioned earlier the Cayley determinant $\binom{0}{0}$ is much less likely to cause numerical instabilities. However the single step of recurrence does not reduce the dimension. Thus we can not recurse all the way down to standard 4-dimensional scalar integrals and we have to extend our integral basis.

The new basis elements are higher dimensional index-less boxes of the form

$$I_4^{[2l],s} \quad : \quad I_4^{[8],s}, \quad I_4^{[6],s}, \quad I_4^{[4],s}, \quad I_4^{[2],s}. \quad (3.57)$$

Due to the presence of the dimensional regulator ϵ in (3.55), it is necessary to know the pole of the right hand side even if we are interested only in the finite part of the answer.

The question of stable evaluation of the new basis integrals will be discussed in the next chapter.

3.3.4. Lower point functions

The methods described in the previous subsection work equally well for triangles and bubbles. Nevertheless there are few subtle points, which should be noted.

The recurrence scheme of type (3.55) can be used to treat small Gram determinants only for IR finite 3-point functions. For the IR divergent triangles it does not work because $\binom{0}{0}_3 = 0$. In that case the left hand side of (3.55) is zero and we get

$$I_3^{[2(l+1)]} = \frac{1}{(d + 2l - \sum_{i=1}^{3} \nu_i + 1)} \sum_{s=1}^{3} \frac{\binom{0s}{0j}_3}{\binom{j}{0}_3} \mathbf{s}^- I_3^{[2l]} \qquad (3.58)$$

where j is a free index, which should be chosen such that $\binom{j}{0}_3 \neq 0$.

The dimension in (3.58) is reduced faster than the indices, which causes the appearance of 2-dimensional bubbles in the reduction.

For 2-point functions, Gram and Cayley determinants have the form

$$()_2 = -2p_1^2 \qquad \binom{0}{0}_2 = 4m_1^2 m_2^2 - (m_1^2 + m_2^2 - p_1^2)^2$$

therefore they can be zero simultaneously if $m_1 = m_2$.

While it would be still possible to use recurrence relations for 2-point functions, due to their simplicity it is advantageous to directly evaluate them for the case $p_1^2 = 0$. See the complete list of special cases for 1- and 2-point functions in generic dimension in Appendix C.

4. Numerical code PJFry

In any Feynman diagram based computation eventually we face the problem of getting accurate numerical values for the amplitude. And hence the numerical values for tensor integrals.

One possibility is to reduce all tensor integrals to a 4-dimensional scalar integral basis analytically which is successfully used for processes with up to four external states. But with increasing number of particles the growing complexity of the phase-space confronts us with many exceptional configurations which are challenging for the numerical evaluation of tensor integrals.

In addition to the classic Passarino-Veltman approach there are several alternative methods available for multileg processes. All of them suffer from numerical cancellations in the different regions of the same physical phase space. Therefore it is impossible to choose one best scheme on a *per process* basis. This takes analytic reduction off the table and forces us to resort to numerical reduction where we can choose the scheme on a *per point* basis. In this approach we treat tensor integrals as the basic amplitude building blocks in the analytic part of the calculation. Then expressions are numerically evaluated with help of a library for tensor integral coefficients, which can select appropriate reduction algorithm *on the fly*.

The complete automation of one-loop corrections greatly benefits from publicly available tools. They allow code reuse, testing, independent comparisons and further improvement by the scientific community.

Currently there are several publicly available codes relevant to tensor integral evaluation.

The 4-dimensional scalar integrals with up to four external legs, which are used as basic building blocks in many reduction algorithms have two complete numeric implementations

- QCDLoop/FF package [87, 170] — all finite and divergent scalar 4-dimensional integrals with real masses in dimensional regularization.

- OneLOop [167] — all finite and divergent scalar 4-dimensional integrals with real or complex masses in dimensional regularization.

For the tensor integrals the situation is more complicated

- LoopTools/FF [110, 113, 170] – tensor integrals up to rank 4 with up to 5 legs. No soft-collinear case. No treatment of small four point Gram determinants.

- Golem95C [41, 62] – tensor integrals up to rank 6 with up to 6 legs. Numerical integration for small Gram determinants in massless case. Massive is potentially unstable for small four point Gram determinants. Complex internal masses.

As one can see in case of tensor integrals there is no single publicly available library, which would handle all physically relevant mass combinations equally well.

Our aim is to fill the gap and provide fast and stable public implementation of tensor reduction suitable for any physically relevant kinematics.

4.1. Program description

The PJFry is an opensource library for the numerical evaluation of one loop tensor integrals. Is is licensed under the GNU Lesser General Public License. The latest version can be obtained from the program web-page at https://github.com/Vayu/PJFry/wiki.

The core functionality is written in ANSI/ISO C++ and can be compiled with any supporting compiler (e.g. g++ or icpc). The program needs an external library for the evaluation of 4-dimensional scalar integrals. Currently QCDLoop [87] and OneLOop [167] are supported. The interface to C, C++,

Fortran 77, Fortran 95 and Mathematica is provided (see more details and examples in the corresponding sections).

We implement the dimensional recurrence algorithms described in the previous chapter. Due to the recursive nature of the algorithms, one can greatly benefit from reusing building blocks throughout the calculation. This is done by implementing a cache system.

The tensor 5-point functions are reduced down to the basis of 4-dimensional boxes, triangles, bubbles and tadpoles in several steps.

According to Sec. 3.3.2 pentagon tensor form-factors can be reduced without introduction of inverse Gram determinants $()_5$. We express them in terms of 4-point integrals in shifted dimension and powers of denominators as shown in Fig. 3.2. These four-point integrals also enter (3.20) as form-factors of tensor boxes.

For each of the the five 4-point subkinematics the $I_{4,ij...}^{[2l],s}$ we choose the optimal reduction scheme depending on the value of the corresponding 4-point Gram determinant:

for $s = 1 \to 5$ **do**
    if $\binom{s}{s}_5 \ll 1$ **then** {small Gram determinant}
        reduce to $I_4^{[2l],s}$ + triangles with (B.2)
    else
        reduce to I_4^s + triangles with (B.1)
    end if
end for

The reduction of 3-point integrals $I_{3,ij...}^{[2l],st}$, coming from the reduction of boxes and form-factors of tensor triangles, is performed similarly with an additional check for IR finiteness

for $s, t = 1 \to 5$ **do**
    if $\binom{st}{st}_5 \ll 1$ **then** {small Gram determinant}
        if $\binom{0st}{0st}_5 = 0$ **then** {IR divergent}
            reduce to $I_3^{[2l],st}$ + bubbles with (B.5)
        else {IR finite}

                reduce to $I_3^{[2l],st}$ + bubbles with (B.4)
            end if
        else
            reduce to I_3^{st} + bubbles with (B.3)
        end if
    end for

Finally the off-shell bubbles are reduced with recurrence (B.6) and explicit analytic representations are used for on-shell bubbles (see App. C).

4.2. Evaluation of scalar integrals

The reduction scheme outlined above gives us an answer in terms of scalar one-loop integrals. In case of a large Gram determinant they are 4-dimensional and can be evaluated directly by the underlying scalar integral library.

For the small Gram determinant case we have to extend our basis with higher dimensional integrals as explained in Sec. 3.3.3:

$$\begin{aligned} I_4^{[2l],s} &: I_4^{[8],s}, \quad I_4^{[6],s}, \quad I_4^{[4],s}, \quad I_4^{[2],s}, \\ I_3^{[2l],st} &: I_3^{[8],st}, \quad I_3^{[6],st}, \quad I_3^{[4],st}, \quad I_3^{[2],st}. \end{aligned} \quad (4.1)$$

There are several options for the numerical evaluation of the additional basis integrals (4.1):

1. Analytic expressions in terms of polylogarithms are not known so far. While it is in principle possible to obtain them it would require a tremendous amount of work. Each mass-kinematic configuration is a separate non-trivial calculation. Another drawback is the fact that the resulting expressions are likely to suffer stability problems in small Gram region due to cancellations between different logarithmic terms.

2. Numeric integration of Feynman parameter integrals. The idea is to derive efficient one-dimensional Feynman parameter representations for integrals in shifted dimension, which behave well in the small Gram re-

gion. The resulting expressions are integrated numerically. This approach has implemented for the massless case in Golem95 program [41, 62]. But in the massive case it is much more difficult to derive one-dimensional Feynman representations due to the increased complexity and larger amount of possible configurations.

3. Series expansion in the small Gram region [94]. This method treats all mass-kinematic combinations uniformly with dimensional recurrence, therefore works equally well for both massive and massless cases. It has better efficiency than the numerical integration method and is successfully used in massless [103] and massive [75] calculations.

In our program we employ the third method – series expansion by dimensional recursion. To derive the expansion formula we start with (3.23) and write it for the case of index-less integrals in the following form

$$X(d+2l-n+1)I_n^{[2(l+1)]} = I_n^{[2l]} - Z_n^{[2l]} \qquad (4.2)$$

where

$$X = \frac{()_n}{\binom{0}{0}_n} \quad \text{and} \quad Z_n^{[2l]} = \sum_{s=1}^{n} \frac{\binom{s}{0}_n}{\binom{0}{0}_n} I_n^{[2l],s}$$

Now depending on the value of X this formula contains three distinct cases:

1. $X \gg 1$ — regular downward recursion to 4-dimensional I_n and lower point functions.

2. $X = 0$ — $I_n^{[2l]}$ degenerates into the sum of $n-1$ point functions $Z_n^{[2l]}$.

3. $X \ll 1$ — upward recursion leading to power series expansion in X.

For the purpose of this section we are interested in the small X upward recursion. For the first few steps we get

$$I_n^{[2l]} = Z_4^{[2l]} + X(d+2l-n+1)I_n^{[2(l+1)]}$$
$$= Z_n^{[2l]} + X(d+2l-n+1)\left(Z_n^{[2(l+1)]} + X(d+2(l+1)-n+1)I_n^{[2(l+2)]}\right)$$
$$\vdots$$

$$I_n^{[2l]} = \sum_{m=0}^{M} a_m^{(l)} X^m Z_n^{[2(l+m)]} + \overbrace{\left[a_M^{(l)} X^M I_n^{[2(l+M)]}\right]}^{\text{series remainder}}, \quad a_m^{(l)} = 2^m \left(\frac{d+2l-n+1}{2}\right)_m \tag{4.3}$$

where $(a)_m = \Gamma(m+a)/\Gamma(a)$ is the Pochhammer symbol.

This type of series expansion was extensively studied in [97]. For our purposes it is enough to know that the series converges for $X \ll 1$, so by cutting it at some M we will get an approximation to $I_n^{[2l]}$ in terms of $Z_n^{[2l]}$ which contains sums of lower point functions.

For the evaluation of $n-1$ point functions $I_n^{[2l],s}$ entering $Z_n^{[2l]}$ we can use (4.2) again in one of the three regimes, depending on the value of new X.

The different algorithms used in the program are schematically illustrated for tensor pentagon form factors of ranks 2 and 3 in Fig. 4.1. The blue lines correspond to the reduction formulae derived in Section 3.3.2.

In the reduction of 4-point functions teal lines represent direct downward recursion for large Gram determinant (3.1), while the solid red path is an alternative scheme with Cayley determinant (B.2). Relations which are shared by both schemes are drawn with two-colored dashed lines. The small $()_4$ series expansion procedure (4.3) discussed in this section is depicted for $M=3$ by dotted red lines. Similarly for 3-point functions we have green lines for (B.3), magenta for (B.4) of $I_{3,i}^{[2]}$. And dotted magenta lines for the series expansion of scalar $I_3^{[2l]}$ in small $()_3$ case.

4.2.1. Evaluation of UV poles

One important feature of the expansion (4.3) is that coefficients $a_m^{(l)}$ depend on the dimension d. The higher dimensional integrals are generally UV-divergent,

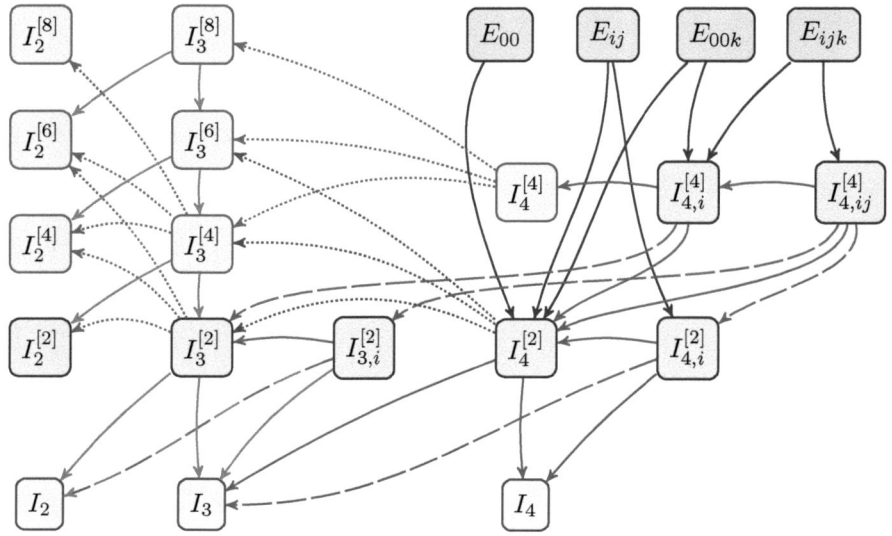

Figure 4.1.: Rank 3 pentagon form-factors calculation flowchart

therefore the pole part of $Z_n^{[2l]}$ will contribute to the finite term of the $I_n^{[2l]}$.

Alternatively we can take advantage of exact analytic expressions for UV-poles of $I_n^{[2l]}$ and rewrite (4.3) for the finite term of ϵ expansion

$$\bar{I}_n^{[2l]} = \sum_{m=0}^{M} \bar{a}_m^{(l)} X^m \left(Z_n^{[2(l+m)]} + b_m^{(l)} D_n^{[2(l+m)]} \right) + R_M(X^M)$$
$$\bar{a}_m^{(l)} = 2^m \left(\frac{5+2l-n}{2} \right)_m \qquad b_m^{(l)} = -\sum_{i=1}^{m} \frac{2}{3-n+2(l+i)} \quad (4.4)$$

where $R_M(X^M)$ is the remainder of the truncated series expansion. And we introduced a notation for the finite $\bar{I}_n^{[2l]}$ and UV-divergent $D_n^{[2l]}$ parts of the ϵ-expansion for $I_n^{[2l]}$:

$$I_n^{[2l]} = \bar{I}_n^{[2l]} + \frac{1}{\epsilon} D_n^{[2l]} + O(\epsilon) \quad (4.5)$$

From simple power counting in (3.21) it is evident that integrals $I_n^{[2l]}$ are UV-divergent if $l \geq n-2$. The explicit expressions for the poles $D_n^{[2l]}$ can be obtained by different methods. For example by extracting them from the expansion of a hypergeometric representation [97]. On the other hand, one can directly use recurrence (3.23) starting with known poles of 4-dimensional tadpoles and bubbles. The Gram determinant always cancels from the final expression for the pole.

From (C.1) we can immediately write down the pole of $I_1^{[2l]}$:

$$D_1^{[2l]} = \frac{(-1)^l}{(2l+2)!!} Y_{11}^{l+1} \qquad (4.6)$$

where Y_{ij} is the Cayley matrix element (3.4) and the double factorial is defined as

$$(2n-1)!! = \prod_{i=1}^n (2i-1) = \frac{(2n)!}{n!\, 2^n}. \qquad (4.7)$$

Using dimensional recurrence and canceling denominators we get for bubbles

$$D_2^{[0]} = 1 \qquad (4.8)$$

$$D_2^{[2]} = -\frac{1}{6}(Y_{11} + Y_{12} + Y_{22}) \qquad (4.9)$$

$$D_2^{[4]} = \frac{1}{120}\left(3Y_{11}^2 + 3Y_{11}Y_{12} + 2Y_{12}^2 + Y_{11}Y_{22} + 3Y_{12}Y_{22} + 3Y_{22}^2\right) \qquad (4.10)$$

$$D_2^{[6]} = \frac{1}{1680}\Big(5Y_{11}^3 + 5Y_{12}Y_{11}^2 + Y_{22}Y_{11}^2 + 4Y_{12}^2 Y_{11} + Y_{22}^2 Y_{11}$$
$$+ 3Y_{12}Y_{22}Y_{11} + 2Y_{12}^3 + 5Y_{22}^3 + 5Y_{12}Y_{22}^2 + 4Y_{12}^2 Y_{22}\Big) \qquad (4.11)$$

One can continue this series as well as apply it to the divergent parts of the triangles and bubbles [94].

The $D_n^{[2l]}$ can be compactly written for $l \geq n-2$ and $1 \leq n \leq 4$ in the following form

$$D_n^{[2l]} = \frac{(-1)^l}{(2l+3-n)!} \sum_{i_1}\sum_{i_2}\cdots\sum_{i_{2L-1}}\sum_{i_{2L}} \left(Y_{[i_1 i_2}Y_{i_3 i_4}\cdots Y_{i_{2L-1} i_{2L}]}\right), \quad L = l-n+2 \qquad (4.12)$$

where the $2L$-nested summation is done from 1 to n such that $i_{m+1} \geq i_m$ for any m. And the square brackets denote the $(2L-1)!!$ non-equivalent distributions of $2L$ indices over L matrices Y (as in (3.9)).

The formula (4.12) can be generalized to the case of $D_{n,jk...}^{[2l]}$ with the help of the $\nu_{ji_1...i_{2L}}$ symbols (3.22).

4.2.2. Convergence acceleration

Evaluating $Z_n^{[2l]}$ in (4.3) for large l can be quite expensive. Therefore it is beneficial to apply convergence acceleration methods, which will allow to achieve the target numerical accuracy with fewer terms in the expansion. The factorial growth of the coefficients (4.3) suggests that methods developed for asymptotic series would be appropriate.

The Wynn epsilon method [151, 176] is a member of a large family of similar so-called lozenge (or rhombus) transformations [174]. The algorithm is defined as a non-linear recurrence relation:

$$\epsilon_{-1}^{(i)} = 0, \tag{4.13a}$$

$$\epsilon_0^{(i)} = S_i, \tag{4.13b}$$

$$\epsilon_{k+1}^{(i)} = \epsilon_{k-1}^{(i+1)} + 1/\left(\epsilon_k^{(i+1)} - \epsilon_k^{(i)}\right). \tag{4.13c}$$

Where S_i is the sequence of partial sums of either (4.3) or (4.4)

$$S_i = \sum_{m=0}^{i} a_m^{(l)} X^m Z_n^{[2(l+m)]} \quad \bar{S}_i = \sum_{m=0}^{i} \bar{a}_m^{(l)} X^m \left(Z_n^{[2(l+m)]} + b_m^{(l)} D_n^{[2(l+m)]}\right) \tag{4.14}$$

The computed values of $\epsilon_k^{(i)}$ can be arranged in a table

$$\begin{array}{cccc} \epsilon_0^{(0)} & \epsilon_1^{(0)} & \epsilon_2^{(0)} & \cdots \\ \epsilon_0^{(1)} & \epsilon_1^{(1)} & \epsilon_2^{(1)} & \cdots \\ \epsilon_0^{(2)} & \epsilon_1^{(2)} & \epsilon_2^{(2)} & \cdots \\ \vdots & \vdots & \vdots & \ddots \end{array} \tag{4.15}$$

where the first column are the starting values of the recursion $\epsilon_0^{(i)} = S_i$ according to (4.13b). And the rest of the table can be computed with relation (4.13c) which connects values located at the vertices of a rhombus:

$$\begin{matrix} & \epsilon_k^{(i)} & \epsilon_{k+1}^{(i)} \\ \epsilon_{k-1}^{(i+1)} & \epsilon_k^{(i+1)} & \end{matrix} \qquad (4.16)$$

Thus by starting with a sequence of M partial sums the rules (4.13) will allow us to evaluate all elements $\epsilon_j^{(m-j)}$ with $0 \leq m \leq M$ and $0 \leq j \leq m$.

From the viewpoint of mathematical analysis the ϵ-algorithm is a transformation of the partial sums of the series into its Padé approximants:

$$\epsilon_{2k}^{(i)} = [i+k/k]_f(x), \qquad i,k \in \mathbb{N} \qquad (4.17)$$

and the best approximation for the M-element sequence of partial sums will be given by

$$\epsilon_{2\nu}^{(M-2\nu)} = [\nu/\nu], \qquad \nu = \lfloor M/2 \rfloor. \qquad (4.18)$$

In our program we have used the Wynn epsilon method to improve the convergence of the series expansion of $I_4^{[2l]}$ and $I_3^{[2l]}$ in the region of small Gram determinant.

4.3. Interface and usage examples

The source code of the PJFry tensor reduction program can be obtained in prepackaged form from the download section of the project web-page https://github.com/Vayu/PJFry/downloads. Alternatively the latest development version can be taken directly from Git code repository by running

```
> git clone git://github.com/Vayu/PJFry.git
```

To ensure portability and simplify the installation process the program uses the GNU build system (also known as Autotools). The distribution package is supplied with a configuration script, so the basic installation is done by running the sequence of commands

```
> tar xjf pjfry-1.x.y.tar.bz2
> cd pjfry-1.x.y
> ./configure
> make
> make check
> sudo make install
```

This will compile and install to `/usr/local` the dynamic library `libpjfry.so` and the Mathematica MathLink interface `PJFry` (assuming the user has the Mathematica system and the configuration script has found it).

Additional advanced configuration options are explained in the provided `INSTALL` file. Here we just list them:

```
--enable-golem-mode
--enable-f2c
--with-mcc-path=/path/mcc
--with-integrals
```

To facilitate compatibility with existing software, we decided to adopt Loop-Tools [113] conventions for the library main interface. This allows switching between PJFry and LoopTools in the same calculation with minimal changes.

The conventions for the momenta and masses labeling are shown in Figure 4.2. It should be noted that they slightly differ from conventions of Chapter 3, which our program uses internally. The difference is in the cyclic shift of mass labels and change of sign of the chords q_i.

The external kinematics for a $n \leq 5$ point loop integral can be specified with $n(n-1)/2$ Lorentz-invariant quantities. We choose them to be n external masses p_i^2 and $n(n-3)/2$ generalized Mandelstam variables s_{ij} (3.3). Which with n internal masses gives total of $n(n+1)/2$ variables.

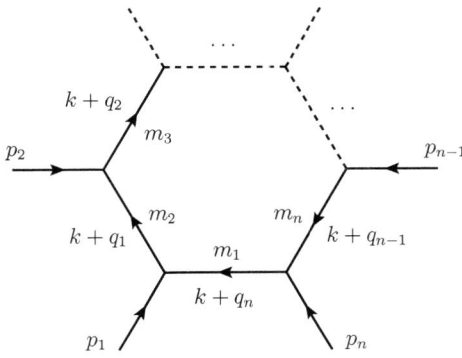

Figure 4.2.: All-incoming momenta labeling

This gives us for scalar integrals

$$I_5 = E_0(\{p_i\}, \{m_i\})$$
$$= E_0(p_1^2, p_2^2, p_3^2, p_4^2, p_5^2, s_{12}, s_{23}, s_{34}, s_{45}, s_{15}, m_1^2, m_2^2, m_3^2, m_4^2, m_5^2) \quad (4.19\text{a})$$
$$I_4 = D_0(\{p_i\}, \{m_i\}) = D_0(p_1^2, p_2^2, p_3^2, p_4^2, s_{12}, s_{23}, m_1^2, m_2^2, m_3^2, m_4^2) \quad (4.19\text{b})$$
$$I_3 = C_0(\{p_i\}, \{m_i\}) = C_0(p_1^2, p_2^2, p_3^2, m_1^2, m_2^2, m_3^2) \quad (4.19\text{c})$$
$$I_2 = B_0(\{p_i\}, \{m_i\}) = B_0(p_1^2, m_1^2, m_2^2) \quad (4.19\text{d})$$
$$I_1 = A_0(\{m_i\}) = A_0(m_1) \quad (4.19\text{e})$$

The main library interface for the languages C and C++ is defined in the header file `pjfry.h`. The C++ interface is a set of global functions encapsulated inside the `PJFry` namespace. The C interface is compatible with FORTRAN calling conventions and therefore there is no need for a separate FORTRAN interface.

There are individual functions for form-factors of every rank $R \leq n$ of an n-point tensor integral with $n \leq 5$. The names of these functions are summarized in the Table 4.1. The functions take tensor indices and kinematic invariants as arguments and return the complex value of the corresponding tensor form-factor.

Rank	0	1	2	3	4	5
Pentagons	E0v0	E0v1	E0v2	E0v3	E0v4	E0v5
Boxes	D0v0	D0v1	D0v2	D0v3	D0v4	
Triangles	C0v0	C0v1	C0v2	C0v3		
Bubbles	B0v0	B0v1	B0v2			
Tadpole	A0v0					

Table 4.1.: Tensor form-factor functions' names

Two additional functions control the evaluation of scalar integrals which depend on the regularization scale μ_R. The square of this scale can be set by calling `SetMu2(muRsq)` and will be later passed on to the underlying scalar integrals library. The current value of the scale can be obtained by calling `GetMu2()`. By default $\mu_R = 1$ at the start of the program.

4.3.1. Cache system

Any Feynman diagram calculation can be characterized by the highest tensor rank which appears in it. The tensor reduction of higher rank tensors shares a lot of common structures with the lower rank tensors. Moreover some of their building blocks map directly on the lower point tensors.

Consider our example for pentagon rank 2 and 3 form-factors in Figure 4.1. By evaluating all dependencies for $E_{00}, E_{ij}, E_{00k}, E_{ijk}$ we get the following structures as a by-product $D_i^s, D_{00}^s, D_{ij}^s, D_{00k}^s, C_i^{st}, C_{00}^{st}, B_{00}^{stu}$ for all subkinematics of the original 5-point kinematics (according to (3.20)). Those lower point functions are marked by nodes with green background. The nodes marked with red background are needed only in case of small Gram determinant and still can be shared among all other form-factors.

As we can see for non-exceptional kinematics the reduction is done in terms of lower point integrals. Ideally after reducing rank 3 pentagons we would need to add only $I_{4,ijk}^{[6],s}$ to get the complete set of rank 3 boxes.

That recursive nature of the reduction scheme can be reflected in the program code by demanding to organize the calculation in a certain restricted top-down way. One would start with highest point kinematics in the process and evaluate all tensor form-factors for this kinematics and for every subkinematics with pinched lines in one go. Then repeat for the next highest point configuration left, until everything is evaluated. This allows sharing common structures in the calculation in a very simple and straightforward fashion at cost of putting restrictions on the way the library is used. A very similar approach is implemented in the Golem95 reduction program [41, 62].

This is a natural way to organize the tensor reduction and we use it in the core routines of PJFry. Therefore the compatibility with Golem95 is very easily archived and there is a separate low-level interface for that.While allowing maximum computational performance the strict top-down approach is a bit more difficult to use in practice. One might find it easier to think of tensor form-factors just as functions of Mandelstam variables, without going into details of their intricate relationships. Such more user friendly interface can be obtained by adding caching system to the program.

The aim of the caching system implemented in the PJFry library is to give maximum transparency and flexibility at minimal performance cost. This allows us to lift some of the restrictions of the top-down scheme although not all of them.

The caching system allows to reuse *identical*, but not *related* objects which might appear in the calculation. For instance tensor integrals are invariant over shifts in loop momentum k. Therefore form-factors of original and shifted integrals are related to each other through rotational identities. Implementing these identities in the cache system would result in *permanent* performance hit, which contradicts our goal. On the other hand without these identities we would still get the result for both integrals, while retaining the possibility to get maximum performance by transforming them into each other analytically.

Internally basic objects which contain the reduction algorithms are the instances of subclasses of the `MinorBase` class declared in `minor.h`. For example `Minor5` provides means to evaluate any signed minor or integral in the reduction tree originated by its 5-point kinematics. When constructed `Minor5` will add itself to the 5-point kinematics cache as well as every child-kinematics 4-, 3- and 2-point caches. Therefore all requests of lower point objects will be redirected by the cache system to the exiting `Minor5` object.

Unfortunately it is very difficult to reconstruct a higher point `Minor` object, from a possibly incomplete set of available cached lower point `Minor`'s. Thus it is highly advisable to *start* an evaluation with top-level kinematics. This way the cache will be populated with links to all child configurations and further evaluations can proceed efficiently in any order.

Due to the important role which they play in the caching system we list here all possible subkinematics of 5-point kinematics *(k5)* explicitly. There are five 4-point, ten 3-point and ten 2-point subkinematics. We use the notation of (4.19) with lower indices denoting pinched lines in Fig. 4.2. To avoid cluttering of the expressions we will often omit squares on the masses and external momenta throughout this chapter. It is assumed that $\mathtt{m1} \equiv m_1^2$ and $\mathtt{p1} \equiv p_1^2$, and so on.

$(k5) = $ p1,p2,p3,p4,p5,s12,s23,s34,s45,s15,m1,m2,m3,m4,m5
$(k5)_1 = $ s12,p3,p4,p5,s45,s34,m1,m3,m4,m5
$(k5)_2 = $ p1,s23,p4,p5,s45,s15,m1,m2,m4,m5
$(k5)_3 = $ p1,p2,s34,p5,s12,s15,m1,m2,m3,m5
$(k5)_4 = $ p1,p2,p3,s45,s12,s23,m1,m2,m3,m4
$(k5)_5 = $ p2,p3,p4,s15,s23,s34,m2,m3,m4,m5

$(k5)_{12} = $ s45,p4, p5, m1, m4, m5 $\qquad$ $(k5)_{123} = $ p5, m1, m5
$(k5)_{13} = $ s12,s34,p5, m1, m3, m5 $\qquad$ $(k5)_{124} = $ s45, m1, m4
$(k5)_{14} = $ s12,p3, s45,m1, m3, m4 $\qquad$ $(k5)_{125} = $ p4, m4, m5
$(k5)_{15} = $ p3, p4, s34,m3, m4, m5 $\qquad$ $(k5)_{134} = $ s12, m1, m3
$(k5)_{23} = $ p1, s15,p5, m1, m2, m5 $\qquad$ $(k5)_{135} = $ s34, m3, m5
$(k5)_{24} = $ p1, s23,s45,m1, m2, m4 $\qquad$ $(k5)_{145} = $ p3, m3, m4
$(k5)_{25} = $ s23,p4, s15,m2, m4, m5 $\qquad$ $(k5)_{234} = $ p1, m1, m2
$(k5)_{34} = $ p1, p2, s12,m1, m2, m3 $\qquad$ $(k5)_{235} = $ s15, m2, m5
$(k5)_{35} = $ p2, s34,s15,m2, m3, m5 $\qquad$ $(k5)_{245} = $ s23, m2, m4
$(k5)_{45} = $ p2, p3, s23,m2, m3, m4 $\qquad$ $(k5)_{345} = $ p2, m2, m3

The last function of the public PJFry interface is the function `ClearCache`. It has no computational overhead because it just marks the cache as empty. To archive the best performance it is recommended to call this function before every new phase-space point in the calculation. This way old entries will not slowdown cache look ups.

4.3.2. Mathematica interface

If the configuration script detects the installation of the Wolfram Mathematica computer algebra system, the MathLink interface to PJFry library will be build during compilation. This interface mimics the C++ interface and we will use it to provide concrete examples of library usage. We assume $\mu_R = 1$ in this section unless specified otherwise.

We start by listing all functions with a brief description:

`Install["PJFry"]` starts the PJFry MathLink connection.

`Names["PJFry`*"]` prints the list of exported functions.

`GetMu2[]` returns the current value of the regularization scale squared.

`SetMu2[musq]` sets the regularization scale squared to the new value `musq`.

`ClearCache[]` empties internal caches.

`E0v0[p1,p2,p3,p4,p5,s12,s23,s34,s45,s15,m1,m2,m3,m4,m5,ep]`
returns the value of a scalar 5-point function (ep $\in$ 0, 1, 2 term of ϵ-expansion).

all 5-point form-factors take 5-point kinematics as part of their argument lists (as explicitly shown above for scalar `E0v0`). For brevity we abbreviate it as "...`k5`...". The last parameter `ep` $\in$ 0, 1, 2, which selects term of ϵ-expansion, is optional (default is 0 – finite part).

E0v1[i,...k5...,ep] returns the ith tensor form-factor of the vector 5-point function

E0v2[i,j,...k5...,ep] returns the ij^{th} form-factor of the rank-2 tensor 5-point function

E0v3[i,j,k,...k5...,ep] returns the ijk^{th} form-factor of the rank-3 tensor 5-point function

E0v4[i,j,k,l,...k5...,ep] returns the $ijkl^{\text{th}}$ form-factor of the rank-4 tensor 5-point function

E0v5[i,j,k,l,m,...k5...,ep] returns the $ijklm^{\text{th}}$ form-factor of the rank-5 tensor 5-point function

D0v0[p1,p2,p3,p4,s12,s23,m1,m2,m3,m4,ep] returns the value of the scalar 4-point function (ep $\in 0,1,2$ term of ϵ-expansion).

all 4-point form-factors take 4-point kinematics as part of their argument lists (as explicitly shown above for scalar **D0v0**). For brevity we abbreviate it as "...k4...". The last parameter ep $\in 0,1,2$, which selects term of ϵ-expansion, is optional (default is 0 – finite part).

D0v1[i,...k4...,ep] returns the ith tensor form-factor of the vector 4-point function

D0v2[i,j,...k4...,ep] returns the ij^{th} form-factor of the rank-2 tensor 4-point function

D0v3[i,j,k,...k4...,ep] returns the ijk^{th} form-factor of the rank-3 tensor 4-point function

D0v4[i,j,k,l,...k4...,ep] returns the $ijkl^{\text{th}}$ form-factor of the rank-4 tensor 4-point function

The triangles and bubbles have similar syntax.

By default lower point form-factors can only be evaluated for subkinematics of previously evaluated pentagons and boxes. If a triangle or bubble is not found in cache the program will return `Indeterminate`. It is done intentionally to encourage proper utilization of the cache system. This behavior can be changed by adding `#define USE_TRIANGLES "1"` to the `common.h` header file, it is also enabled by default for Golem-like interface.

The example Mathematica session, which connects to PJFry and evaluates scalar 5-point function and two vector triangles:

```
In:=  Install["PJFry"]
      PJFry MathLink
      Type Names["PJFry`*"] to show exported names
Out=  LinkObject["PJFry", 5, 5]

In:=  E0v0[0,0,5,0,5,121,-33.5,34.2,35.8,-30.6,0,0,0,5,5]
Out=  0.0001297601422591693 + 0.0002759474276598341i

In:=  C0v1[1,35.8,0,5,0,5,5]
Out=  0.00227375738344195 + 0.05105477571330563i

In:=  C0v1[1,5,35.8,0,5,0,5]
Out=  Indeterminate
```

as we can see the second triangle returned `Indeterminate` since it wasn't subkinematics of the 5-point function and the `USE_TRIANGLES` option is not enabled.

We can evaluate $1/\epsilon$ and $1/\epsilon^2$ coefficients by supplying the last parameter which is assumed to be zero by default (0 for finite part, 1 for single pole and 2 for double pole):

```
In:=  E0v0[0,0,5,0,5,121,-33.5,34.2,35.8,-30.6,0,0,0,5,5, 1]
Out=  7.18504605754102 · 10⁻⁶ − 0.00006571739937150293i

In:=  E0v0[0,0,5,0,5,121,-33.5,34.2,35.8,-30.6,0,0,0,5,5, 2]
Out=  −1.414528240005221 · 10⁻⁶ + 0i
```

For the next example we consider the benchmark phase-space point from section 10.4.1 of [25]. In that kinematic configuration the 4-point Gram

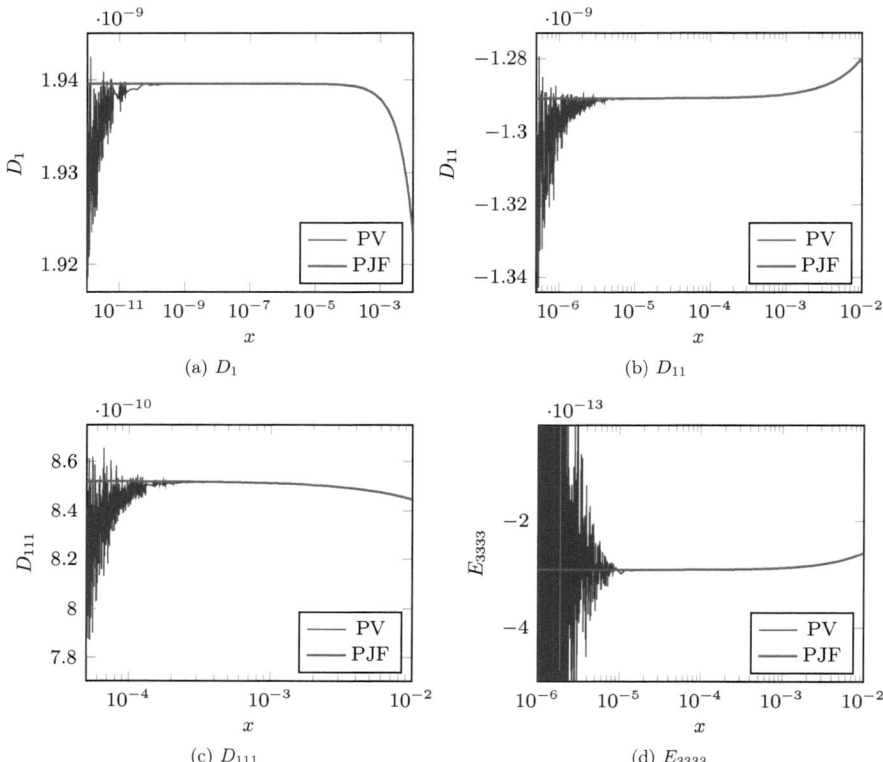

Figure 4.3.: Box and pentagon form-factors with simple Passarino-Veltman reduction and PJFry code in small Gram region.

determinant becomes zero, thus spoiling the numerical accuracy of tensor reduction.

This box kinematics is defined as

$$D_{...}(s_{45}, t_{16}(x), 0, 0, t_{23}, s_{345}, 0, 0, 0, M^2)$$

where

$$s_{45} = 1 \cdot 10^4, \qquad s_{345} = 2 \cdot 10^4, \qquad t_{23} = -4 \cdot 10^4, \qquad M = 91.1876,$$

and
$$t_{16} = t_{\text{crit}}(1+x) = \frac{s_{345}(s_{345} - s_{45} + t_{23})}{s_{345} - s_{45}}.$$

The Gram determinant is linearly proportional to x and goes to zero as $x \to 0$:

$$()_4 = -2s_{345}\, t_{23}\, (s_{345} - s_{45} + t_{23})\, x$$

Additionally we define the pentagon kinematics, which contains the aforementioned 4-point configuration as subkinematics. This will allow us to see how errors propagate from lower point functions to higher point functions in the reduction.

$$E_{...}(0, 0, t_{16}(x), 0, 0, s_{45}, t_{234}, s_{345}, t_{23}, s_{34}, 0, m^2, 0, 0, M^2)$$
$$t_{234} = -3.5 \cdot 10^4, \qquad s_{34} = 1.5 \cdot 10^4, \qquad m = 80.938.$$

We have chosen coefficients D_1, D_{11}, D_{111} and E_{3333} as representatives of their respective tensor rank. For other coefficients the picture is essentially the same and we omit them for simplicity. A separate version of the reduction library with the simple Passarino-Veltman reduction was prepared for comparisons. Alternatively one may use the LoopTools reduction program [113], which is possible because this configuration has no double soft-collinear singularities.

We compare results of the simple Passarino-Veltman reduction with those from the PJFry library, which implements additional reduction schemes for the case of small Gram determinants. In Fig. 4.3 a selection of tensor formfactors is plotted against the dimensionless variable x approaching the special point $x = 0$. It is evident from the plots, that the straightforward Passarino-Veltman reduction (blue line) becomes more and more numerically unstable while approaching zero Gram determinant point. Note that we are not plotting the same range on all plots, because the level of instability is worse for higher rank coefficients. It reflects the fact that each rank reduction step introduces an additional inverse power of the Gram determinant. On the other hand the PJFry code (red line) looks smooth and stable. In the plotted region PJFry is already using expansion algorithms and the accuracy is actually increasing towards $x = 0$.

As expected the instability from the box form-factors is propagating to the E_{3333} pentagon coefficient (Fig. 4.3d), thus confirming that the accurate reduction of 4-point functions is crucial for the evaluation of higher point loop integrals. The instability of E_{3333} is slightly lower than in highest contributing box due to the damping effect of the reduction coefficient.

We can further investigate the accuracy of the expansion algorithms in PJFry by taking advantage of two observations:

1. The Passarino-Veltman approach works well for large Gram determinant (large x).

2. The integral at $x = 0$ degenerates into the sum of lower point functions (4.3) and therefore is known exactly.

After establishing reliable points at $x \gg 1$ and $x = 0$ we can use interpolation techniques to approximate form-factors between these values. Then we will use the interpolated expression as a reference to estimate the accuracy of the expansion algorithms.

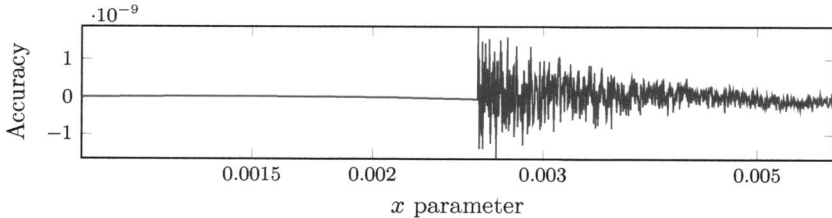

Figure 4.4.: Relative error estimate for the E_{3333} form-factor around the point where PJFry switches to small Gram expansion.

We have performed this procedure for the E_{3333} form-factor using a 6-degree polynomial interpolation. The result of comparison of PJFry output with interpolated values is shown in Fig. 4.4. The plotted region includes the point at which PJFry switches from Passarino-Veltman reduction to the Gram-expansion mode. The smooth line in the left half of the plot corresponds to the small Gram expansion mode. The accuracy is very high and the plot resolution does not allow to see actual deviations from the interpolated expression, which are of order 10^{-14}. The right half shows the increasing instability of standard reduction reaching $\approx 10^{-9}$ relative deviation at the switching point.

Knowing that PJFry gives more than 9 correct digits (at least in this configuration) we can use it as a reference in the next plot. In Fig. 4.5 we show the relative difference between a simple Passarino-Veltman reduction and PJFry for box coefficients of three different ranks. It is clear that the instabilities of a simple reduction in the small Gram region are unbound and expressing a power growth $\delta \sim (1/x)^R$.

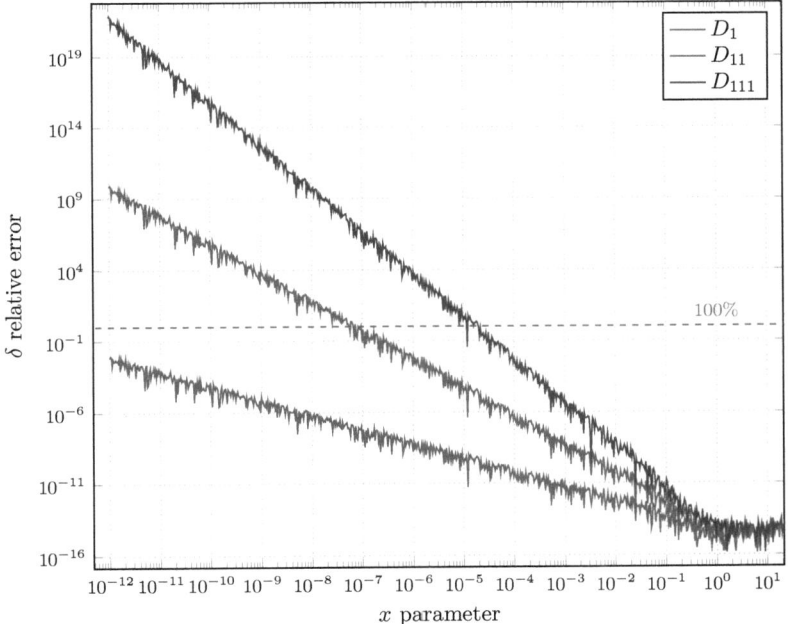

Figure 4.5.: Relative error of Passarino-Veltman reduction for Gram determinant approaching zero.

One might argue that it would be possible to get better accuracy in plot 4.4 by switching to the expansion mode earlier. And indeed we can push it quite a bit by changing the switching point by hand. Unfortunately there is no general algorithm which for a given phase-space point could chose the most accurate reduction scheme. Therefore we have to resort to the heuristic estimates of "badness" of the point. For example in case of small Gram determinant such characteristic can be

$$\chi = S \frac{()_4}{\binom{0}{0}_4},$$

where S is some scale with dimension of squared mass introduced to make χ dimensionless. For instance we can use the maximum element of the Cayley matrix as $S = \max(Y)$. By selecting some χ_{cut} we can categorize all phase-space points into $\chi < \chi_{\text{cut}}$ and $\chi \geq \chi_{\text{cut}}$. The optimal value for χ_{cut} depends on many factors and can be found only experimentally.

In PJFry we have several such variables responsible for the selection of different reduction algorithms. The values of these cutoffs have been chosen based on a series of test runs. Despite being sometimes a bit conservative they usually provide an accuracy sufficient for practical applications. Nevertheless the advanced user might decide to adjust them to adapt to some particularly tricky kinematic configuration. This can be done by editing the constant definitions in `minor.cpp` file.

4.4. Fully contracted tensor test

The relations between tensor integrals (3.14) which are used in Passarino-Veltman reduction can be used to test the correctness of the program and to estimate the numerical accuracy.

We contract tensor integral with various external momenta and use (3.13) to cancel propagators on the right hand side. Thus obtaining a relation between n-point tensor integral of rank R and integrals with lower rank and number of legs.

For example by multiplying a rank 1 pentagon by $2q_1$ we reduce it to the sum of scalar pentagon and scalar boxes with some kinematic coefficients:

$$2q_{1,\mu}I_5^\mu = (m_2^2 - m_1^2 - q_1^2)I_5 + I_4^{(1)} - I_4^{(5)} \tag{4.20}$$

substituting the form-factor expansion

$$2p_1^2 E_1 + (s_{12} - p_2^2 + p_1^2)E_2(s_{45} - s_{23} - p_1^2)E_3 + (p_5^2 - s_{15} + p_1^2)E_4 =$$
$$= (m_2^2 - m_1^2 - p_1^2)E_0 + D_0^{(1)} - D_0^{(5)} \tag{4.21}$$

The same procedure applied to Rank 2 pentagon multiplied by $2q_1q_2$:

$$4q_{1,\mu}q_{2,\nu}I_5^{\mu\nu} = (m_2^2 - m_1^2 - q_1^2)(m_3^2 - m_1^2 - q_2^2)I_5 + (m_2^2 - m_1^2 - q_1^2)I_4^{(2)} + I_3^{(12)}$$
$$+ (m_3^2 - m_1^2 - q_2^2)I_4^{(1)} - 2q_{2,\mu}I_4^{\mu,(5)} - (m_2^2 - m_1^2 - q_1^2 - 2q_1q_2)I_4^{(5)} - I_3^{(15)} \tag{4.22}$$

substituting form-factor expansion

$$2(s_{12} - p_2^2 + p_1^2)E_{00} + 2p_1^2(s_{12} - p_2^2 + p_1^2)\bar{E}_{11} + 2s_{12}(s_{12} - p_2^2 + p_1^2)E_{22} +$$
$$+ ((p_1^2 - p_2^2)^2 + s_{12}(6p_1^2 - 2p_2^2 + s_{12}))E_{12} + 74 \text{ terms} =$$
$$= (m_2^2 - m_1^2 - q_1^2)(m_3^2 - m_1^2 - q_2^2)E_0 + (m_2^2 - m_1^2 - q_1^2)D_0^{(2)} +$$
$$+ (m_3^2 - m_1^2 - q_2^2)D_0^{(1)} + (p_1^2 - s_{12} - p_2^2)D_1^{(5)} + (p_1^2 - s_{45} + p_3^2 - p_2^2)D_2^{(5)} +$$
$$+ (p_1^2 + s_{34} - p_5^2 - p_2^2)D_3^{(5)} + C_0^{(12)} - C_0^{(15)} \tag{4.23}$$

One can independently evaluate left and right hand sides of (4.21) and (4.23) numerically. Comparing resulting values gives an estimate of the accuracy loss difference between more complex left hand side objects and simpler right hand side objects.

There are four independent vectors in 5-point kinematics, which form a basis in 4-dimensional Minkowski space-time. Therefore it is sufficient to check only contractions of pentagons with $q_1, \ldots, q_4$, because any other 4-tensor can be expressed as a linear combination of products of those.

For accuracy tests of the 5-point tensor integrals we selected the following set of 16 contractions:

Rank 1: $I_5(q_1)$, $I_5(q_2)$, $I_5(q_3)$, $I_5(q_4)$
Rank 2: $I_5(q_1, q_2)$, $I_5(q_1, q_3)$, $I_5(q_1, q_4)$, $I_5(q_2, q_3)$, $I_5(q_2, q_4)$, $I_5(q_3, q_4)$
Rank 3: $I_5(q_1, q_2, q_3)$, $I_5(q_1, q_2, q_4)$, $I_5(q_1, q_3, q_4)$, $I_5(q_2, q_3, q_4)$ \hfill (4.24)
Rank 4: $I_5(q_1, q_2, q_3, q_4)$
Rank 5: $I_5(q_1, q_2, q_3, q_4, q_1)$

where $I_5(q_1, q_4)$ should be thought as $I_5^{\mu\nu} q_{1,\mu} q_{4,\nu}$.

The expressions size grows rapidly with increasing tensor rank, as we can see from (4.21) and (4.23). The expansion of a rank 5 pentagon in terms of 80 tensor form-factors and products of external momenta contains ≈ 25000 terms, so it is necessary to use computer algebra to generate the test identities. A subroutine for the FORM algebra system [172] has been written to generate contraction identities.

For general kinematics the 16 identities (4.24) are quite lengthy and take 50000 lines of C code. Such big expressions inevitably lead to spurious numerical cancellations and precision loss. We used them only to check the general correctness of the code at the limited number of random phase-space points.

In order to estimate the accuracy of tensor reduction and to avoid unnecessary cancellations it is better to generate specialized test identities for the particular kinematic configuration. This allows us to take into account various symmetries like equal particle masses or zero masses. This analytic simplification step reduces the code size considerably, especially in case of high rank tensors (by several orders of magnitude).

For test purposes in this section we have chosen five kinematic configurations:

- The first is a completely massless pentagon, encountered in jet production processes in QCD (see also Sec. 4.5). It is the simplest 5-point kinematics but nevertheless not trivial, there is a possibility to have small Gram determinants inside the physical phase-space after cuts, which are potentially problematic for the naïve Passarino-Veltman approach. Due to the fact that internal masses are fixed to zero, massless kinematics allows many analytic optimization and simplifications. Therefore, it will be interesting to see how completely generic algorithm performs in such specialized configuration.

- The other four correspond to massive pentagons of jet-associated heavy quark pair production. This configuration has one internal mass and therefore has much more different small Gram determinant regions inside the physical phase-space. Therefore it is a good check of different reduction schemes and algorithms implemented in the program.

Comparing massless and massive cases will allow us to estimate the impact of internal masses on the program accuracy.

4.4.1. Massless configuration

For the massless configurations we have performed several tests with 10^8 randomly generated phase-space points each. We excluded soft and collinear regions of phase-space where evaluating tensor integrals numerically would serve no practical purpose. This was done by applying a simple cut on the kinematic variables:

$$p_i \cdot p_j < E_{\text{cms}} \delta_{\text{cut}} \qquad \delta_{\text{cut}} = 10^{-11}, \ 10^{-3} \qquad (4.25)$$

The accuracy plots are shown in Figs. 4.6 and 4.7. It is worth to note that the value of δ_{cut} does not significantly affect the accuracy in this test.

4.4.2. Massive configuration

The four massive pentagon diagrams appearing in jet associated heavy quark pair production are shown in Fig. 4.8. We label them by the number of integral massive lines as C1m, C2m, C3m and C4m. Those diagrams contain pentagon tensor integrals up to rank 4. In principle only integrals of maximum rank 3

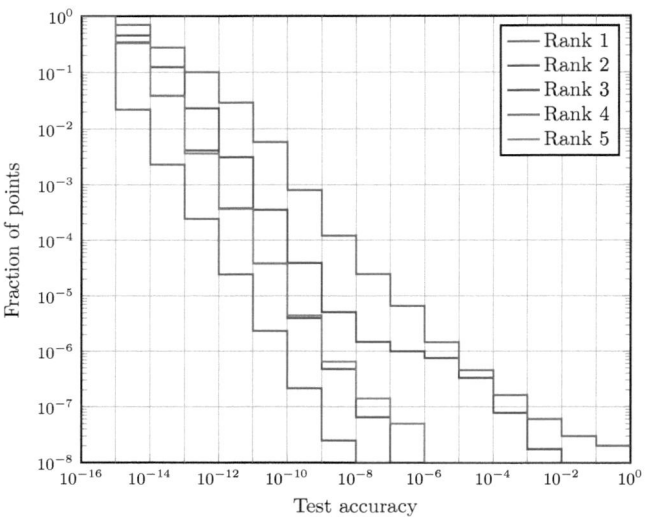

Figure 4.6.: Fraction of points with relative test accuracy less than a value. Massless kinematics, 10^8 points in flat phase-space, $\delta_{\text{cut}} = 10^{-11}$.

would be needed in practical calculation, because one tensor rank can always be canceled.

We evaluate all our selected contraction tests (4.24) for each of four configurations, with kinematic cut $\delta_{\text{cut}} = 10^{-7}$ and for three different center-of-mass energies. The results are shown in Figs. 4.9-4.11. We do not show C3m and C4m for $m = 3m$ because they look identical to ones with $m = 4m$.

The approximate speed of the program measured in these tests is 3 ms for the evaluation of all tensor form-factors up to rank 5 on the 3.40 GHz Intel Xeon CPU.

Despite being useful in verifying general program correctness and identifying potential problems the synthetic contraction tests give quite limited estimate of the final amplitude precision. Tensor integrals enter amplitude calculation with kinematic coefficients, which can reweight the errors and change relative importance of different terms. Additionally there might be unanticipated large cancellations between terms reducing overall available precision. To estimate

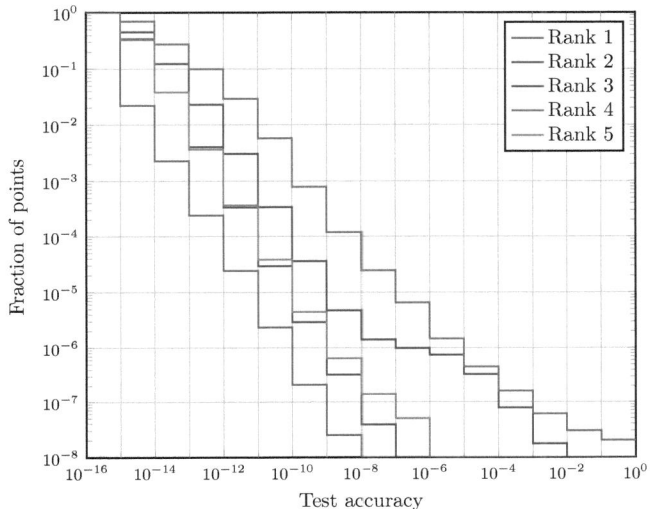

Figure 4.7.: Fraction of points with relative test accuracy less than a value. Massless kinematics, 10^8 points in flat phase-space, $\delta_{\text{cut}} = 10^{-3}$.

the impact of these factors we will do some amplitude calculations using our tensor reduction program.

4.5. Five gluon amplitude test

In this section we compute helicity amplitudes for the five gluon scattering subprocess of 3 jet production. It is the most challenging $2 \to 3$ process for a Feynman diagrammatic approach for a several reasons. It contains the highest rank tensor integral for 5-point processes in renormalizable theories. The large number of diagrams and the simple form of the answer known from string-based calculations [28] suggests the presence of big gauge cancellations between diagrams and therefore potential numerical problems.

Of course this is not a test of tensor reduction alone, but of the interplay of various parts of the calculation which combine together into an amplitude. The accuracy of such calculation depends on how efficiently we do diagrams'

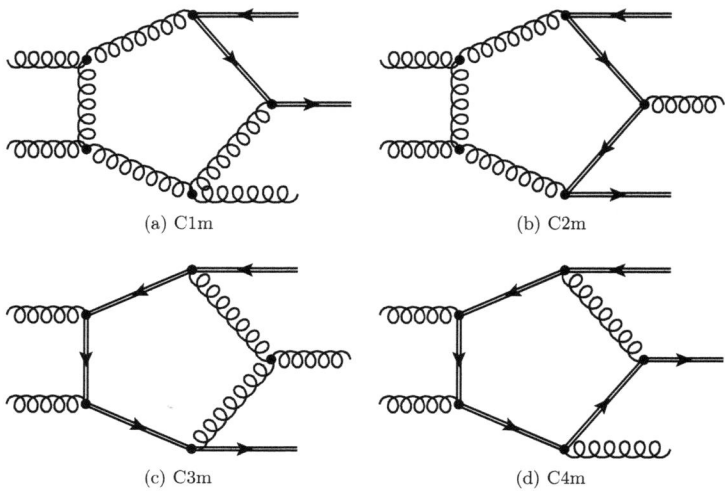

Figure 4.8.: Massive pentagons in $t\bar{t} + j$

processing and how many simplifications can be found analytically. We should keep this in mind when we interpret results of the test.

To compare results and estimate the accuracy we will use the NGluon package [13]. It has an option of evaluating amplitude with higher precision using the QD package [117], which is very useful for accuracy comparisons.

4.5.1. Calculation method

Following the decomposition of [28] we calculate the single independent color-ordered configuration, while the rest can be obtained via decoupling equations [26]. From 711 diagrams contributing to the one-loop five gluon amplitude only 208 contribute to the leading color.

For this calculation we employed the spinor-helicity method described in [109, 124, 132]. The external gluons are assigned specific helicities and their polarizations are expressed in terms of massless spinors and one arbitrary

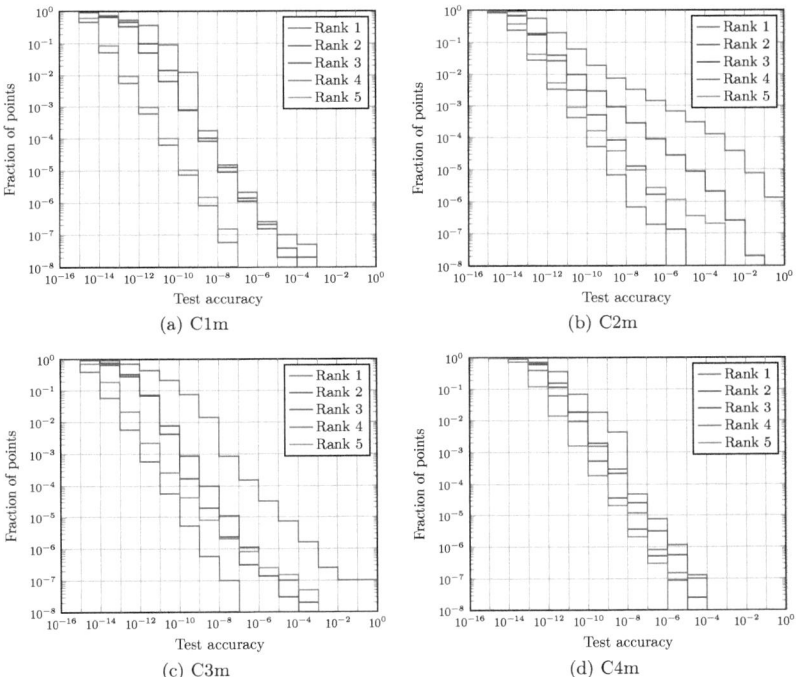

Figure 4.9.: Fraction of points with relative test accuracy less than a value. Massive kinematics, 10^8 points in flat phase-space, $E_{\text{CMS}} = 2.2m$.

massless reference momentum k:

$$\epsilon_\mu^+(p_i, k_i) = \frac{[p_i|\gamma_\mu|k_i\rangle}{\sqrt{2}\langle k_i|p_i\rangle}, \qquad \epsilon_\mu^-(p_i, k_i) = -\frac{\langle p_i|\gamma_\mu|k_i]}{\sqrt{2}[k_i|p_i]}. \qquad (4.26)$$

There are three independent helicity configurations in five gluon scattering amplitude:

$$A_5^{(1)}(+++--), \qquad A_5^{(1)}(++++-), \qquad A_5^{(1)}(+++++), \qquad (4.27)$$

and the others can be expressed via permutation and helicity inversion relations.

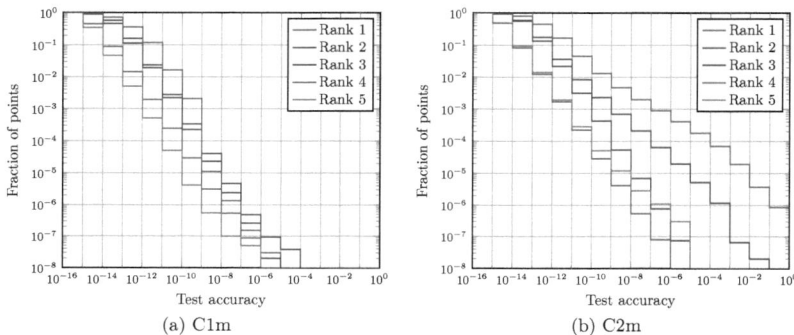

Figure 4.10.: Fraction of points with relative test accuracy less than a value. Massive kinematics, 10^8 points in flat phase-space, $E_{\text{CMS}} = 3m$.

The diagrams have been processed in FORM [172] in order to generate the expressions for each of three basis amplitudes (4.27). After substituting tensor integral form-factor expansions (3.8) the amplitudes are written in form of tensor form-factors with coefficients expressed in terms of spinor products $\langle ij \rangle$ and $[ij]$. Which up to a phase are equal to square roots of Mandelstam variables $s_{ij} = (p_i + p_j)^2$.

The (4.26) representation of polarization vectors is equivalent to the choice of axial gauge. The amplitude is independent of reference vectors k_i and their specific choice can considerably simplify the expressions.

Our choice of reference vectors is:

$$+ + + - - \quad k_1 = k_2 = k_3 = p_5, \quad k_4 = k_5 = p_3 \quad \text{(4.28a)}$$
$$+ + + + - \quad k_1 = k_2 = k_3 = k_4 = p_5, \quad k_5 = p_3 \quad \text{(4.28b)}$$
$$+ + + + + \quad k_1 = k_2 = k_3 = k_4 = k_5 = \xi \quad \text{(4.28c)}$$

One should keep in mind that presence of the preferred direction ξ in (4.28c) might cause numerical cancellations in certain phase-space regions.

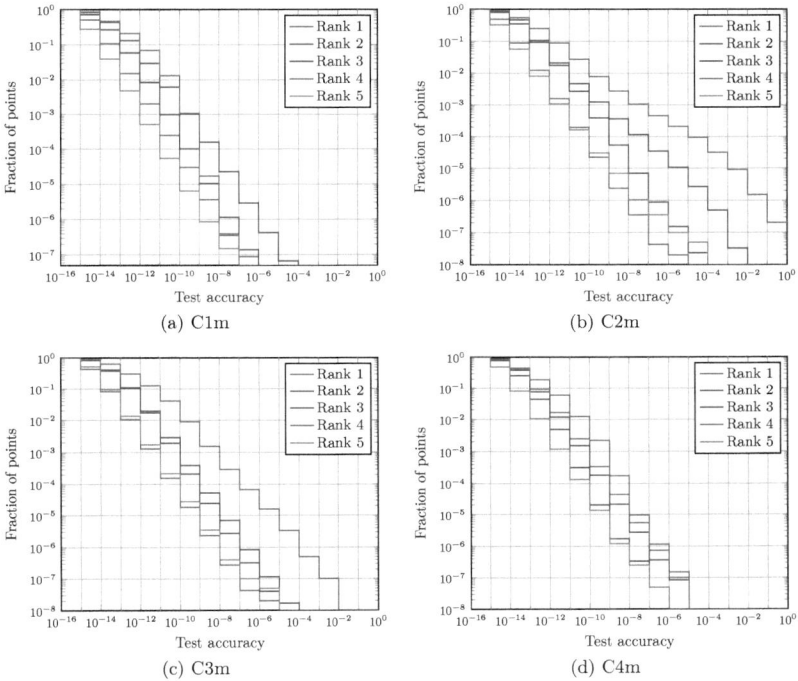

Figure 4.11.: Fraction of points with relative test accuracy less than a value. Massive kinematics, 10^8 points in flat phase-space, $E_{\text{CMS}} = 4m$.

4.5.2. Accuracy plots

The generated amplitudes have been compared to the output of the NGluon package in quadruple precision. We used 5 million points, which were randomly generated in the flat phase-space with a cut (4.25). The accuracy plots for two different values of δ_{cut} are shown in Figs. 4.12 and 4.13.

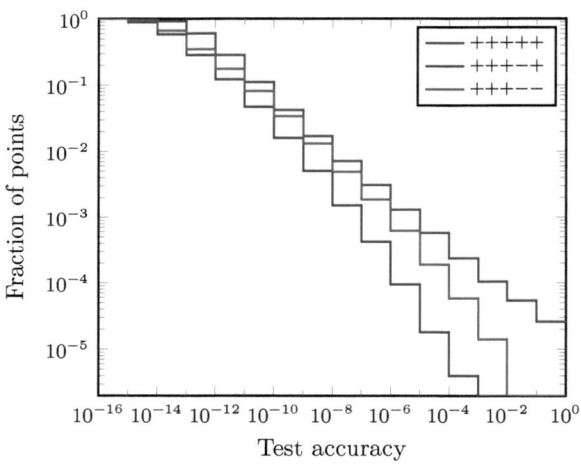

Figure 4.12.: Fraction of points with relative test accuracy less than a value. Five gluon amplitude, $5 \cdot 10^6$ random points, $\delta_{\text{cut}} = 10^{-7}$.

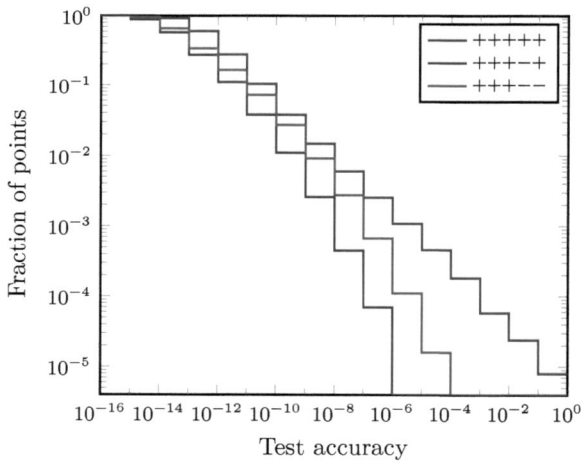

Figure 4.13.: Fraction of points with relative test accuracy less than a value. Five gluon amplitude, $5 \cdot 10^6$ random points, $\delta_{\text{cut}} = 10^{-3}$.

5. QED virtual corrections to the process $e^+e^- \to \mu^+\mu^-\gamma$

In this chapter we present the calculation of QED one-loop virtual corrections to the process $e^+e^- \to \mu^+\mu^-\gamma$ with full mass dependence.

The muon pair production with associated hard photon in electron-positron annihilation was first calculated at tree level by [15, 19]. The first virtual corrections, although with integrated photon variables have been presented in [119]. Shortly after, the completely differential NLO cross section for the same corrections was evaluated in small mass limit [20]. The quasi-collinear logarithmic contributions have been extensively studied in [7, 8, 10, 11, 129].

The first completely differential cross-sections with mass dependence were calculated by two different groups [121, 125]. They included a subset of the full one-loop virtual corrections, which is called ISR and FSR. The ISR stands for Initial State Radiation and corresponds to both loop correction and photon emission applied to the incoming fermion. Likewise FSR – Final State Radiation does the same for the outgoing particle. So far only the ISR+FSR parts of one-loop corrections have been implemented in Monte-Carlo event generators [120, 150].

The calculation of differential cross-sections for the full one-loop virtual corrections have been reported recently [3]. The full corrections have additional contributions from 5-point and associated 4-point functions, which is missing in previous ISR+FSR calculations.

The muon pair production with real photon emission ($e^+e^- \to \mu^+\mu^-\gamma$) is an important background and normalization reaction in the measurement of the pion form-factor:

$$R_{exp} = \frac{\sigma(e^+e^- \to \pi\pi\gamma)}{\sigma(e^+e^- \to \mu^+\mu^-\gamma)} \qquad (5.1)$$

which is necessary for an accurate determination of the anomalous magnetic moment of the muon $(g-2)_\mu$ – one of the most precise tests of the Standard Model [1, 51]. The pion form-factor is a non-perturbative quantity and cannot be reliably calculated within perturbative QCD, but it can be measured experimentally at high luminosity meson factories such as DAFNE, PEP-II, KEKB and BEPC.

In this respect contributions from previously unconsidered QED 5-point functions may become important for accurate predictions at low energy e^+e^- annihilation experiments like BaBar, KLOE and BES.

This is also an ideal benchmark process for testing a tensor reduction code

- Two different massive particles in the loop and five external legs create a complicate phase-space landscape with many potential Gram determinant instabilities.

- Large differences of scales in the problem (up to 7 orders of magnitude) cause accuracy losses due to numerical cancellations.

- The small electron mass gives rise to quasi-collinear configurations, which are nearly singular.

In addition to that, the relatively small number of diagrams and the absence of hadrons in the initial state allows us to evaluate the virtual corrections to a very high precision. Thus making it possible to use statistical methods to asses the numerical accuracy of the calculation as a whole.

5.1. Notation and conventions

For computational convenience we consider the process in all-outgoing configuration:

$$0 \to e^-(p_1) + e^+(p_2) + \mu^-(p_3) + \gamma(p_4) + \mu^+(p_5) \qquad (5.2)$$

where p_i are particle momenta, which obey mass shell conditions $p_i^2 = m_i^2$ and momenta conservation $\sum p_i = 0$. The masses of the external fermions are preserved in all parts of the calculation.

In addition to the masses, we choose the following 5 independent Mandelstam variables:

$$s_{14},\ s_{24},\ s_{13},\ s_{35},\ s_{45} \qquad s_{ij} = (p_i + p_j)^2. \qquad (5.3)$$

This set of invariants gives slightly better behaving results in the quasi-collinear limits $p_1 \| p_4$ and $p_2 \| p_4$, compared to the standard cyclic choice.

Other s_{ij} invariants can be expressed in terms of (5.3) as

$$\begin{aligned}
s_{12} &= 2m_e^2 - s_{14} - s_{24} + s_{35} \\
s_{23} &= m_\mu^2 - s_{13} + s_{14} + s_{24} - s_{35} + s_{45} \\
s_{34} &= 2m_\mu^2 + 2m_e^2 - s_{14} - s_{24} - s_{45} \\
s_{15} &= 2m_\mu^2 + m_e^2 - s_{13} + s_{24} - s_{35} \\
s_{25} &= m_\mu^2 + m_e^2 + s_{13} - s_{24} - s_{45}
\end{aligned}$$

Using crossing relations one can show that (5.2) describes the process we want to study

$$e^+(-p_1) + e^-(-p_2) \to \mu^+(p_5) + \mu^-(p_3) + \gamma(p_4). \qquad (5.4)$$

The fully differential unpolarized leading order cross section can be written as

$$d\sigma^{(0)} = \frac{1}{2\sqrt{s(s-4m_e^2)}} \frac{1}{4} \sum |M_{\text{Born}}|^2 \, d\Phi_3 \qquad (5.5)$$

where $s = s_{12}$ is the center-of-mass energy squared and the summation sum is performed over polarizations of all particles. The factor $1/4$ corresponds to averaging over spins of incoming electrons and the three particle phase-space Φ_3 is defined as

$$\Phi_3 = (2\pi)^4 \delta^{(4)} \left(\sum_{i=1}^{5} p_i\right) \prod_{i=3}^{5} \frac{d^3 \mathbf{p}_i}{(2\pi)^3 2E_i} \qquad (5.6)$$

5.2. Computation method

We are calculating the unpolarized cross section using the Feynman diagram method in the Dirac spinor formalism.

The topologies are generated by QGRAF [143] and then dressed with particles and momenta by the DIANA program [165] according to the model description file. The resulting output contains a list of Feynman diagrams in the textual representation, which is defined by the TML markup language script [164]. We define a TML style to generate expressions suitable for further processing with the FORM [172] computer algebra system.

For instance for diagram on Fig. 5.1c we get

```
*--#[ n1:
**** (diagram 1)

Local [Amp1] = SPL(3,1)*SPL(1,2)*
        1*VALL(1,mu4)*VLL(1,p4+p3,mlm)*VALL(1,mu2)*VALL(2,mu1)*
        VAA(-p2-p1,mu1,mu2)
        *SPR(5,1)*SPR(2,2)*POL(4, mu4);

*--#] n1:
```

where `VALL` is the photon-fermion vertex, `VLL` is the fermion propagator and the rest is polarization spinors.

On the next step the diagrams are passed through the FROM script `amplitude.frm`, which substitutes Feynman rules according to the selected model (in this case QED).

In addition some general simplifications can be enabled by setting configuration parameters. This includes

- Gamma algebra identities like $\gamma^\mu \gamma^\nu \gamma^\mu = (2-d)\gamma^\nu$
- Transversality condition $p_1^\mu \epsilon_\mu(p_1) = 0$
- Dirac equation
- Momenta conservation

The resulting expressions are written in the FORM tablebase. We use it as an input in the squaring program which sums the diagrams and multiplies them by the complex conjugated set. The fermion lines are connected by the completeness relation and Dirac traces are taken on them.

5.2.1. Born amplitude

The tree level diagrams contributing to the leading order amplitude $|M_{\text{Born}}|^2$ are shown in Fig. 5.1. The ISR and FSR pairs are independently gauge invariant.

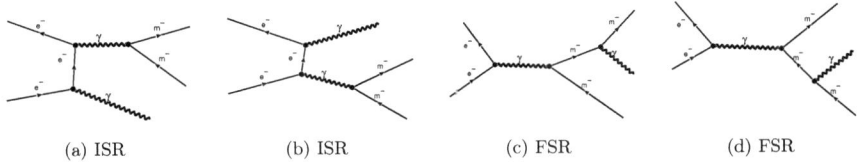

(a) ISR (b) ISR (c) FSR (d) FSR

Figure 5.1.: Tree diagrams for $e^+e^- \to \mu^+\mu^-\gamma$

There is no additional hand-tuning needed at tree level. The leading order squared matrix element obtained in completely automated fashion reads

$$|\mathcal{M}_{\text{Born}}|^2 = (4\pi\alpha)^3 |\mathcal{M}_{\text{Born}}|^2$$

$$|\mathcal{M}_{\text{Born}}|^2 = \frac{16}{s_{12}^2\left(s_{34} - m_\mu^2\right)^2}\Big[m_\mu^4\left(s_{24} - 11m_e^2 + 64s_{14} - 4s_{35} + 4s_{13} - 4s_{45}\right)$$

$$+ m_\mu^2\Big(s_{24}\left(7s_{35} - 6s_{14} - 5s_{45} + 7s_{13} + 17m_e^2\right) + s_{14}\left(4s_{35} + 15m_e^2 + s_{13} - s_{45}\right)$$

$$+ m_e^2\left(6s_{45} - 11s_{35}\right) + 4\left(s_{13} + s_{35}\right)s_{45} - 2s_{35}^2 - 4s_{35}s_{13} - 4s_{13}^2 - 2s_{14}^2 - 6s_{24}^2 - 22m_e^4\Big)$$

$$- \left(s_{14} + s_{24} + s_{45} - 2m_e^2\right)\left(s_{24}^2 + \left(-s_{35} - s_{13} - s_{14}\right)s_{24} + \left(s_{13} - s_{45}\right)s_{14} + m_e^2\left(s_{35} + 3s_{45}\right)\right)\Big] +$$

$$+ \frac{16}{s_{12}^2\left(s_{45} - m_\mu^2\right)^2}\Big[m_\mu^4\left(s_{24} - 7m_e^2 + 4s_{13} - 4s_{45}\right) + m_\mu^2\left(s_{45}\left(-s_{14} - s_{24} + 2m_e^2 + 4s_{13}\right)\right.$$

$$+ s_{14}\left(12m_e^2 - 5s_{24} + 5s_{35} + 5s_{13}\right) + s_{24}\left(4s_{35} + 3s_{13}\right) + m_e^2\left(-9s_{35} + 12s_{24}\right)$$

$$- 2s_{35}^2 - 4s_{35}s_{13} - 4s_{13}^2 - 3s_{14}^2 - 18m_e^4 - 2s_{24}^2\Big) - s_{45}\Big(\left(3m_e^2 - s_{14}\right)s_{45}$$

$$- s_{14}^2 + \left(-s_{24} + s_{35} + 4m_e^2 + s_{13}\right)s_{14} - 6m_e^4 + \left(4s_{24} - s_{35}\right)m_e^2 - s_{24}s_{13}\Big)\Big] +$$

$$+ \frac{16}{s_{12}^2\left(s_{45} - m_\mu^2\right)\left(s_{34} - m_\mu^2\right)}\Big[s_{35}^2\left(-4s_{13} - 16m_e^2 + 6s_{24} + 2m_\mu^2 + 2s_{45} + 4s_{14} - 2s_{35}\right)$$

$$+ s_{35}\Big(s_{24}\left(16s_{13} - 7m_e^2 - 7s_{45} + 15m_\mu^2 - 7s_{14}\right) + m_\mu^2\left(-3s_{14} + 12s_{13} - 4s_{45} + 6m_e^2\right)$$

$$+ s_{14}\left(6s_{13} - 3s_{45} + 9m_e^2\right) + s_{13}\left(4s_{45} - 8m_e^2\right) - 7s_{24}^2 + 2m_e^4 - 4s_{13}^2 - 2s_{14}^2 + 6m_e^2 s_{45} - 12m_e^4 - 2s_{45}^2\Big)$$

$$+ s_{24}^2\left(-7s_{13} - 6m_e^2 + 4s_{14} + 5s_{45} + 6m_\mu^2 + 3s_{24}\right) + s_{24}\Big(\left(9s_{14} - 14s_{13} + 6s_{45} - 15m_e^2\right)m_\mu^2$$

$$+ s_{14}^2 + \left(-2m_e^2 - 8s_{13} + 3s_{45}\right)s_{14} + \left(s_{45} - 2s_{13}\right)\left(2s_{45} - 7m_e^2 - 2s_{13}\right)\Big)$$

$$+ \left(16m_e^2 - 8s_{13} - 2s_{14} + 8s_{45}\right)m_\mu^4 + \left(s_{14}^2 + \left(2s_{45} - 16s_{13} - 9m_e^2\right)s_{14} + 8s_{13}^2\right.$$

$$+ \left(-8s_{45} + 8m_e^2\right)s_{13} + 16m_e^4\Big)m_\mu^2 - \left(s_{14}s_{13} + m_e^2\left(s_{45} - 2s_{13}\right)\right)\left(s_{14} - 4s_{13} + 2s_{45}\right)\Big] \quad (5.7)$$

As additional cross check of our computation procedure (5.7) was compared with a result of FormCalc [113].

5.2.2. Loop amplitude

There are 32 diagrams contributing to $e^+e^- \to \mu^+\mu^-\gamma$ at one-loop level. They can be classified into several independently gauge invariant contributions:

- ISR – photon emission and loop correction on the initial state fermion line,

- FSR – photon emission and loop correction on the final state fermion line,
- Mix – mixed ISR/FSR interference corrections,
- Penta – pentagons and boxes with two different fermion lines inside,
- VP – vacuum polarization correction to virtual photon propagator.

The last group results in the multiplicative correction to the amplitude, which is related to the charge renormalization. This is a universal correction and it can be accounted for in any QED calculation by simply running the fine structure constant to the appropriate scale [44, 73]. Therefore we omit those diagrams from our loop amplitude definition and make substitution

$$\alpha = \alpha(0) \to \alpha(\mu_R^2) = \frac{\alpha}{1 - (\Pi_2(\mu_R^2) - \Pi_2(0))} \quad (5.8)$$

where Π_2 is the full photon propagator.

The self energy corrections to massive external legs are dealt with by the on-shell renormalization procedure (see details in Sec. 5.3).

In this chapter we will concentrate our attention on the new and the most complicated subset "Penta", which contains tensor pentagons up to rank 3.

These classes are summarized in Table 5.1 and a complete list of diagrams is presented in Appendix D.

ISR	FSR	Mix	Penta		
I2IP$^{D.1c}$ I2IA$^{D.1d}$	F2FP$^{D.1a}$ F2FA$^{D.1b}$	I3FP$^{D.2k}$	S5FP$^{D.4a}$	S4FP$^{D.3i}$	S4FA$^{D.3k}$
I3IP$^{D.2c}$ I3IA$^{D.2d}$	F3FP$^{D.2a}$ F3FA$^{D.2b}$	I3FA$^{D.2l}$	O5FP$^{D.4b}$	O4FP$^{D.3j}$	O4FA$^{D.3l}$
I3LP$^{D.2i}$ I3LA$^{D.2j}$	F3LP$^{D.2e}$ F3LA$^{D.2f}$	F3IP$^{D.2g}$	S5IP$^{D.4c}$	S4IP$^{D.3g}$	S4IA$^{D.3e}$
I4LP$^{D.3c}$ I4LA$^{D.3d}$	F4LP$^{D.3a}$ F4LA$^{D.3b}$	F3IA$^{D.2h}$	O5IP$^{D.4d}$	O4IP$^{D.3h}$	O4IA$^{D.3f}$

Table 5.1.: Classes of diagrams for $e^+e^- \to \mu^+\mu^-\gamma$ process

The generated diagrams were processed with FORM in a similar fashion like the tree level diagrams. In the second stage the loop amplitude is multiplied by the conjugated tree amplitude and traces are taken to obtain unpolarized interference amplitude. Which enters the definition of the one-loop cross section

$$d\sigma^{(1)} = \frac{1}{2\sqrt{s(s-4m_e^2)}} \frac{1}{4} \sum 2\,\text{Re}\left(M_{\text{Loop}} M_{\text{Born}}^\dagger\right) d\Phi_3 \tag{5.9}$$

Since taking the Dirac trace is summing away all spinor structure the tensor integrals are left contracted only with external momenta p_i and/or metric tensors $g^{\mu\nu}$. This presents an opportunity of tensor rank reduction by canceling some denominators in loop integrals with (3.13).

Special attention has to be paid to the dimensionality of the numerator. For example d-dimensional loop momenta $k^\mu k^\nu$ can get contracted with the 4-dimensional metric tensor $\bar{g}_{\mu\nu}$ from the trace over gamma matrices. The resulting 4-dimensional $\bar{k} \cdot \bar{k}$ cannot directly cancel d-dimensional $k \cdot k$ in the denominator. This can be fixed by adding and subtracting a $(d-4)$-dimensional part $\tilde{k} \cdot \tilde{k}$ which will result in a compensating term [70]

$$\int \frac{d^d k}{i\pi^{d/2}} \frac{\bar{k} \cdot \bar{k}}{(k^2 - m_1^2)\cdots} = \int \frac{d^d k}{i\pi^{d/2}} \frac{(k \cdot k) - (\tilde{k} \cdot \tilde{k})}{(k^2 - m_1^2)\cdots}$$

$$= \int \frac{d^d k}{i\pi^{d/2}} \frac{1}{\cdots} + \int \frac{d^d k}{i\pi^{d/2}} \frac{m_1^2}{(k^2 - m_1^2)\cdots} - \int \frac{d^d k}{i\pi^{d/2}} \frac{\tilde{k} \cdot \tilde{k}}{(k^2 - m_1^2)\cdots} \tag{5.10}$$

where $k = \bar{k} + \tilde{k}$ and $\bar{k} \cdot \tilde{k} = 0$.

The integrals with $\tilde{k}$ in the numerator as the last term in (5.10) can be mapped to higher dimensional integrals [24, 32, 39]

$$\int \frac{d^d k}{i\pi^{d/2}} \frac{(\tilde{k} \cdot \tilde{k})^l}{((k-q_1)^2 - m_1^2)\cdots((k-q_n)^2 - m_n^2)} = (-1)^l \frac{\Gamma(l-\epsilon)}{\Gamma(1-\epsilon)} \frac{d-4}{2} I_n^{[2l]} \tag{5.11}$$

with $d-4$ multiplying the pole of $I_n^{[2l]}$ it results in a rational term to be added to the amplitude.

After tensor integrals are replaced by form-factor representation (3.8) we are left with loop-tree interference written in terms of form-factors multiplied by kinematic coefficients plus a rational contribution

$$2\,\mathrm{Re}\left(M_{\mathrm{Loop}}M_{\mathrm{Born}}^{\dagger}\right) = \sum_{n=1}^{5}\sum K_{...}^{(n)}F_{...}^{(n)} + K^{(0)} \qquad (5.12)$$

The coefficients $K_{...}^{(0)}$ are functions of Mandelstam variables in the same way as (5.7) is. One important remark is that $K^{(0)}$ is not a complete rational term in the sense of [39, 86]. Part of the rational contribution is hidden inside tensor form-factors $F_{...}^{(n)}$.

After algebraic simplifications the (5.12) is used to generate FORTRAN code which uses the PJFry library for the tensor form-factor evaluation.

5.3. Renormalization and scheme dependence

The loop amplitude contains UV-divergences, which have to be removed by renormalization. Since we omitted vacuum polarization diagrams only mass and wave-function renormalizations are needed [73]. Both of them are contained in the fermion propagator counterterm for the corresponding fermion f

$$\delta_{2,f} = (Z_{2,f} - 1), \qquad \delta_{m,f} = Z_{m,f}, \qquad (5.13)$$
$$-i\Sigma_{\mathrm{ct},f} = (Z_{2,f} - 1)(\slashed{q} - m_f) - Z_{m,f}m_f. \qquad (5.14)$$

Due to its multiplicative nature, the wave-function renormalization is proportional to the tree-level amplitude and can taken into account separately. The mass renormalization on the other hand should be included in the Feynman rules and contributes to four counterterm diagrams which subtract singularities from the diagrams in Fig. D.1.

$$M_{\mathrm{Loop}}^{\mathrm{ren}} = M_{\mathrm{Loop}}^{\mathrm{m.r.}} + (\delta_{2,e} + \delta_{2,\mu})M_{\mathrm{Born}} \qquad (5.15)$$

where the index $m.r.$ means that we already included the mass counterterm in that quantity.

In the on-shell renormalization scheme we have

$$\delta_2 = \delta_m = -\frac{\alpha}{4\pi}\left(\frac{4\pi\mu^2}{m^2}\right)^\epsilon \left(\frac{3}{\epsilon} + 4 + 1_{FDH} + O(\epsilon)\right) \tag{5.16}$$

The value above is calculated in the 't Hooft Veltman (HV) regularization scheme. In the FDH scheme the same quantities would have an additional term which is marked with the index "FDH".

The scheme dependence of the loop amplitude can be obtained in a process independent manner [58, 126]. Using this together with (5.16) we can see that the renormalized amplitude is the same in the HV and FDH schemes.

$$M_{\text{Loop}}^{\text{m.r.},HV} - M_{\text{Loop}}^{\text{m.r.},FDH} = -2\frac{\alpha}{4\pi} M_{\text{Born}} \tag{5.17}$$

$$M_{\text{Loop}}^{\text{ren},HV} - M_{\text{Loop}}^{\text{ren},FDH} = 0 \tag{5.18}$$

5.4. Pole structure and μ_R-dependence

The infrared pole and logarithmic terms can be extracted from the universal behavior of QED (and QCD) one-loop amplitudes [58, 126]. The agreement of the numerical calculation with analytic predictions can serve as an additional cross check of the result.

The general pole factorization formula for the amplitude involving massive fermions:

$$M_{\text{Loop}}^{\text{ren}}(\mu^2) = \mathbb{I}_m(\epsilon, \mu^2) M_{\text{Born}} + M_{\text{Loop}}^{\text{fin}}(\alpha(\mu^2)) + O(\epsilon) \tag{5.19}$$

where all μ-dependence of the $M_{\text{Loop}}^{\text{fin}}$ contribution is contained in the renormalized running coupling $\alpha(\mu)$.

For the renormalized amplitude $e^+e^- \to \mu^+\mu^-\gamma$ we get:

$$\mathbb{I}_m = \frac{\alpha}{4\pi} \frac{(4\pi)^\epsilon}{\Gamma(1-\epsilon)} \left\{ \sum_{\substack{j,k=1 \\ j<k}}^{4} e_j e_k \left(\frac{\mu^2}{|s_{jk}|}\right)^\epsilon \left[\frac{1}{\epsilon}\frac{1}{v_{jk}} \ln\frac{1-v_{jk}}{1-v_{jk}} - \frac{1}{2}\left(\ln^2\frac{m_j^2}{|s_{jk}|} + \ln^2\frac{m_k^2}{|s_{jk}|}\right) \right.\right.$$
$$\left.\left. -\frac{\pi^2}{3} + \frac{1}{v_{jk}}\left(\frac{2}{\epsilon}i\pi - \pi^2\right)\Theta(s_{jk})\right] - \sum_{j=1}^{4}\left[\frac{1}{\epsilon} + \frac{1}{2}\ln\frac{m_j^2}{\mu^2} - 2\right]\right\} \quad (5.20)$$

where e_i is the charge of fermion i in the units of the electron charge and

$$v_{jk} = \sqrt{1 - \frac{4m_j^2 m_k^2}{s_{jk} - m_j^2 m_k^2}}. \quad (5.21)$$

The expression consists of several parts related to the singular structure of the amplitude:

- dimensionally regulated IR singularities manifest as explicit poles in ϵ,
- logarithmic terms $\ln^2 m_j$ and $\ln m_j$ which become singular in the limit of vanishing mass,
- constant terms which arise from the noncommutativity of the limits $\epsilon \to 0$ and $m_j \to 0$ and ensure a smooth transition to massless theory.

The poles and the μ-dependent part:

$$\mathbb{I}_m = \frac{\alpha}{4\pi}\frac{(4\pi)^\epsilon}{\Gamma(1-\epsilon)}\left\{-\left(\frac{\mu^2}{|s_{12}|}\right)^\epsilon \frac{1}{\epsilon}\frac{1}{v_{12}}\ln\frac{1-v_{12}}{1-v_{12}} - \frac{2}{\epsilon} - \ln\frac{m_e^2}{\mu^2} \right. \quad (5.22a)$$

$$-\left(\frac{\mu^2}{|s_{34}|}\right)^\epsilon \frac{1}{\epsilon}\frac{1}{v_{34}}\ln\frac{1-v_{34}}{1-v_{34}} - \frac{2}{\epsilon} - \ln\frac{m_\mu^2}{\mu^2} \quad (5.22b)$$

$$+\left(\frac{\mu^2}{|s_{13}|}\right)^\epsilon \frac{1}{\epsilon}\frac{1}{v_{13}}\ln\frac{1-v_{13}}{1-v_{13}} + \left(\frac{\mu^2}{|s_{24}|}\right)^\epsilon \frac{1}{\epsilon}\frac{1}{v_{24}}\ln\frac{1-v_{24}}{1-v_{24}}$$

$$\left. -\left(\frac{\mu^2}{|s_{14}|}\right)^\epsilon \frac{1}{\epsilon}\frac{1}{v_{14}}\ln\frac{1-v_{14}}{1-v_{14}} - \left(\frac{\mu^2}{|s_{23}|}\right)^\epsilon \frac{1}{\epsilon}\frac{1}{v_{23}}\ln\frac{1-v_{23}}{1-v_{23}}\right\}$$
$$(5.22c)$$

We can identify terms in (5.22) with different gauge independent contributions in the full amplitude (see Tab. 5.1), which allows to check them independently.

- (5.22a) – ISR loop with ISR Born,
- (5.22b) – FSR loop with FSR Born,
- (5.22c) – Penta loop with full Born,
- (5.22a) and (5.22b) – ISR and FSR interference plus Mix loop with full Born.

With the help of (5.22) we have explicitly checked the poles and μ-scale dependence of all subamplitudes and found complete agreement.

5.5. Cross-checks

In gauge theories the results should not depend on the gauge choice. While it is in principle possible to do a calculation in arbitrary gauge, it would be order of magnitude more difficult. For the amplitudes with external gauge bosons it is still possible to check gauge invariance to some extent even after making a specific gauge choice. This could be achieved by using the Ward-Takahashi identity

$$k_\mu \tilde{M}^\mu = 0, \qquad \text{where } M = \epsilon_\mu(k)\tilde{M}^\mu \qquad (5.23)$$

which states that amplitude contracted with the gauge boson momentum instead of its polarization vector is equal to zero. In our process we have one external photon, which we used to verify the Ward identity.

Additionally we have compared our amplitude with the calculation of [2, 3], which using the OPP method, at their published phase-space point

	PJFry	OPP [3]
Born	$5.013964825925060 \cdot 10^{-3}$	$5.013964825924999 \cdot 10^{-3}$
Loop pole	$0.331807389470214 \cdot 10^{-3}$	$0.331807389848555 \cdot 10^{-3}$
Loop finite	$1.554830436628185 \cdot 10^{-3}$	$1.554830457128372 \cdot 10^{-3}$

We observe 8 digits agreement for the virtual part.

The ISR and FSR contributions have been checked against an implementation in the Phokhara Monte-Carlo generator [67, 68, 125, 150]. We have found agreement within 6-12 digits of accuracy, depending on the phase-space point.

5.6. Numerical results

In this section we conclude the investigation of the numerical accuracy of PJFry by evaluating differential distributions for two experimental setups. Both of them are so-called *meson factories* which are characterized by relatively low energy, but very high luminosity. They are designed to produce a large number of mesons allowing accurate measurements of their properties and precision checks of the Standard Model.

The Babar experiment was operating at the SLAC e^+e^- accelerator facility PEP-II until April of 2008. During its runs Babar accumulated 530 fb^{-1} integrated luminosity and the data analysis is still ongoing. Another experiment which we will consider in our study is KLOE [5] at the DAΦNE [140] accelerator. KLOE is operating since 1999. The upgraded detector KLOE-2 is at the moment under construction.

The main differences between the two experiments in the context of photon associated muon pair production are the beam energy and the tagged photon angular cuts. Babar operates at 10.58 GeV center-of-mass energy while at KLOE collisions happen at 1.02 GeV. The photons can be collinear to the beam axis at KLOE, unlike Babar where this region is excluded by kinematic cuts. In both respects, KLOE kinematics is more challenging for loop amplitude evaluations. One might expect numerical problems in the quasi-collinear region. Comparing the error estimates for easy Babar cuts with more difficult KLOE cuts will allow us to estimate inaccuracies due to this special kinematic configuration.

The basic idea is to evaluate differential distributions for several selected variables using a large number of phase-space points. With sufficiently small statistical errors we will be able to see systematic code instabilities in bins where errors do not decrease with increasing number of points. The absence of such bins will serve as an indication that there is no significant accuracy loss in the numerical code for the amplitude.

For phase-space integrations the Monte-Carlo event generator Phokhara 7.0 [66] was used. The histogramming facility of Phokhara was modified to bin weighted events and to allow parallel computational runs. The ability to merge results of parallel runs was crucial for the accumulation of enough statistics in a reasonable amount of time.

The phase-space cuts we used roughly correspond to the experimental setup of the KLOE and BaBar detectors.

	BaBar	KLOE
E_{CMS}	10.56 GeV	1.02 GeV
$E_{\gamma,\min}$	3 GeV	0.02 GeV
θ_γ	20°–138°	0°–15°, 165°–180°
Q^2	0.25–50 GeV^2	0.25–1.06 GeV^2
$\theta_{\mu^\pm}$	40°–140°	50°–130°

Table 5.2.: Phase-space cuts for KLOE and BaBar settings. Q^2 is the invariant mass squared of the muon pair.

The other relevant parameters are taken from the PDG report [141]

$$m_e = 0.5109989 \cdot 10^{-3} \text{ GeV}, \ m_\mu = 0.105658367 \text{ GeV}, \ \alpha = 1/137.03599968. \tag{5.24}$$

We have produced distributions for two types of quantities:

- One is the plain renormalized one-loop cross-section, which contains IR poles and therefore depends on the regularization scale

$$d\sigma^{(1)} = \frac{1}{2\sqrt{s(s-4m_e^2)}} \frac{1}{4} \sum 2 \operatorname{Re}\left(M_{\text{Loop}}^{\text{ren}} M_{\text{Born}}^\dagger\right) d\Phi_3 \tag{5.25}$$

- Another is the above cross-section where we subtracted IR poles together with scale-dependent terms and singular mass logarithms (5.19)

$$d\bar{\sigma}^{(1)} = \frac{1}{2\sqrt{s(s-4m_e^2)}} \frac{1}{4} \sum 2\operatorname{Re}\left(M_{\text{Loop}}^{\text{fin}} M_{\text{Born}}^{\dagger}\right) d\Phi_3 \qquad (5.26)$$

Both cross-sections can be separated into gauge-invariant parts according to Tab. 5.1. Since ISR and FSR contributions have been known for years, we do not consider them separately and concentrate here on the new piece "Penta".

The distributions for Born and Born+Loop cross-sections are shown in Figs. 5.2-5.6 for both σ (5.25) and $\bar{\sigma}$ (5.26). The error bars are below one per mil and are too small to be seen in the plots.

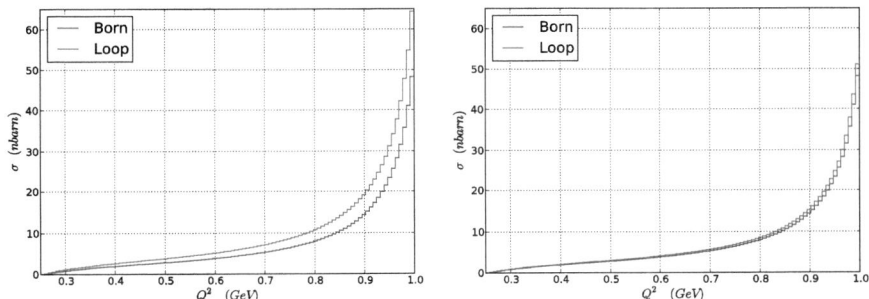

Figure 5.2.: Muon pair invariant mass distribution for KLOE (left σ, right $\bar{\sigma}$).

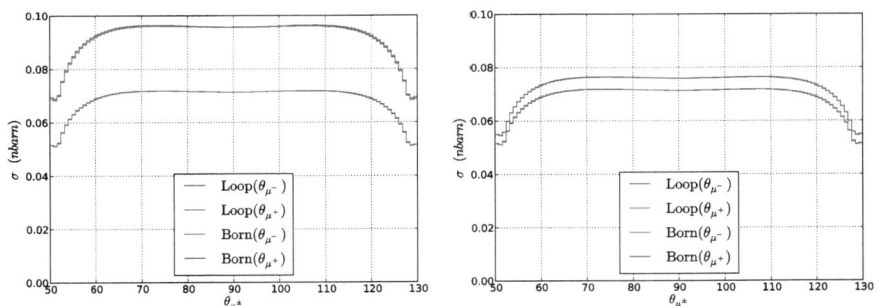

Figure 5.3.: Angular distributions of μ^+ and μ^- for KLOE (left σ, right $\bar{\sigma}$).

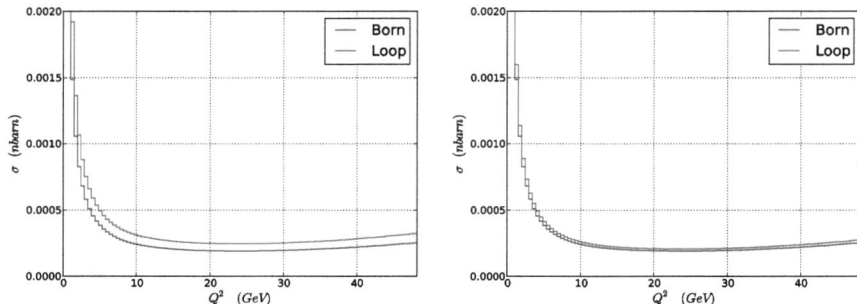

Figure 5.4.: Muon pair invariant mass distribution for BaBar (left σ, right $\bar{\sigma}$).

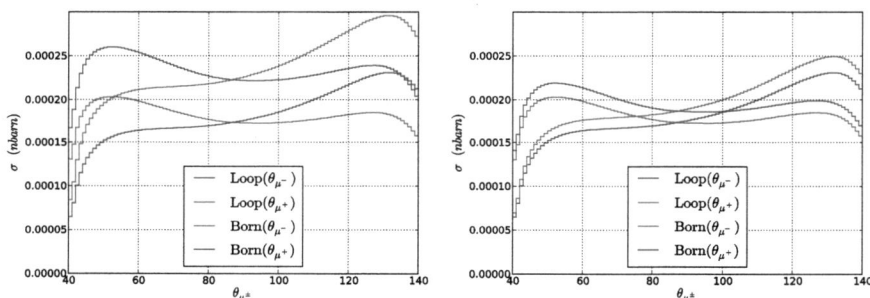

Figure 5.5.: Angular distributions of μ^+ and μ^- for BaBar (left σ, right $\bar{\sigma}$).

Since the pentagon contribution is rather small, it is convenient to plot the ratio of Loop and Born cross-sections, which we define as

$$K = 1 + \sigma^{(1)}/\sigma^{(0)} \qquad (5.27)$$

Then the error can be estimated using

$$\sigma_f^2 = \left|\frac{\partial f}{\partial x}\right|^2 \sigma_x^2 + \left|\frac{\partial f}{\partial y}\right|^2 \sigma_y^2 + 2\frac{\partial f}{\partial x}\frac{\partial f}{\partial y}\text{cov}_{xy}$$

$$f = x/y, \qquad \sigma_f = \frac{1}{y^2}\sqrt{x^2\sigma_y^2 + y^2\sigma_x^2 - 2xy\cancel{\text{cov}_{xy}}}$$

where σ_x and σ_y are standard deviations of the Loop and Born cross-sections

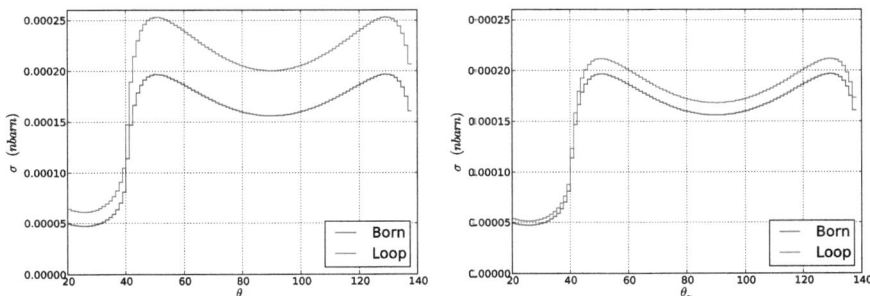

Figure 5.6.: Angular distribution of photon for BaBar (left σ, right $\bar{\sigma}$).

respectively. We assume here that those quantities are not (strongly) correlated, which is an acceptable approximation.

In the plots 5.7-5.13 we show the "Penta" contribution to several frequently used observables. The one standard deviation absolute accuracy estimate is plotted separately below each distribution. Additionally the number of phase-space points used in the Monte-Carlo integration is shown on the each plot.

One can see that there is no visible stability problems in the pentagon evaluation and the errors are dominated by the statistical uncertainty. The evaluation time for the full amplitude per phase-space point is of order 0.3 ms on the 2.80 GHz Intel Xeon CPU.

For comparison we show the relative size of different gauge invariant contributions for the KLOE configuration in Figs. 5.14-5.19. It is evident that pentagons give a noticeable contribution to the shape of angular distributions of muons and their forward-backward asymmetry. The effect is of the order of few per mil and can become important when more accurate measurements become available. This gives a motivation for the inclusion of "Penta" contributions into the Monte-Carlo event generators together with the corresponding real corrections. A similar conclusions can be drawn for the BaBar configuration, although the effect is a bit smaller (see Figs. 5.11, 5.12).

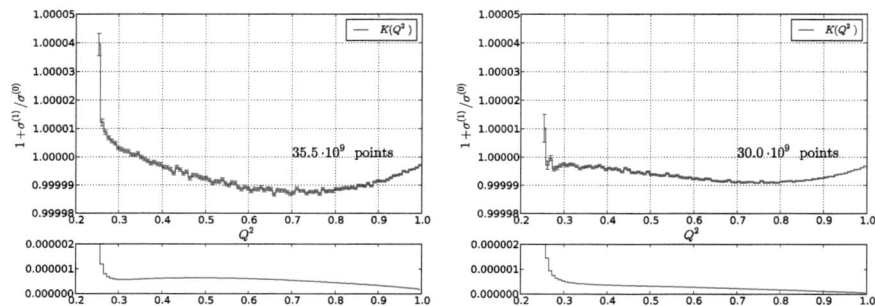

Figure 5.7.: "Penta" contribution to muon pair invariant mass distribution for KLOE setup (left σ, right $\bar{\sigma}$). Bottom: absolute error estimate.

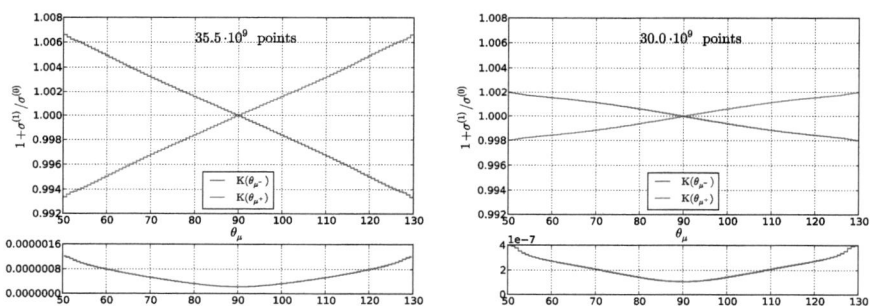

Figure 5.8.: "Penta" contribution to angular distributions of μ^+ and μ^- for KLOE setup (left σ, right $\bar{\sigma}$). Bottom: absolute error estimate.

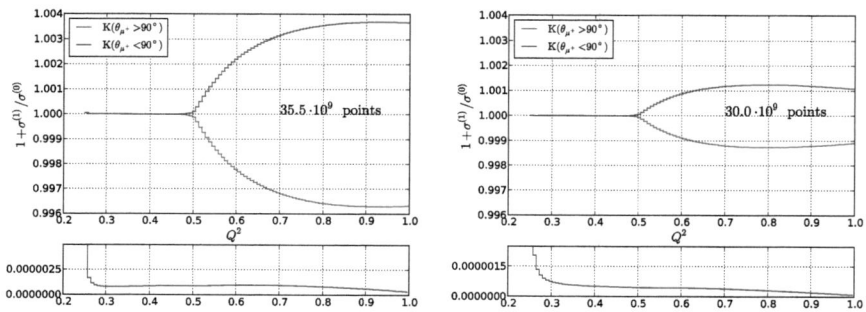

Figure 5.9.: "Penta" contribution to forward-backward asymmetry of μ^+ for KLOE setup (left σ, right $\bar{\sigma}$). Bottom: absolute error estimate.

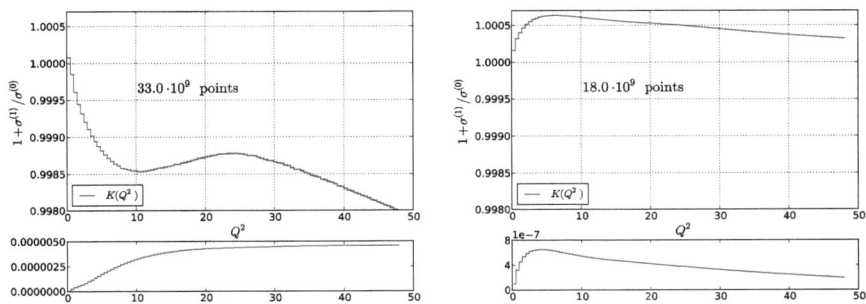

Figure 5.10.: "Penta" contribution to muon pair invariant mass distribution for BaBar setup (left σ, right $\bar{\sigma}$). Bottom: absolute error estimate.

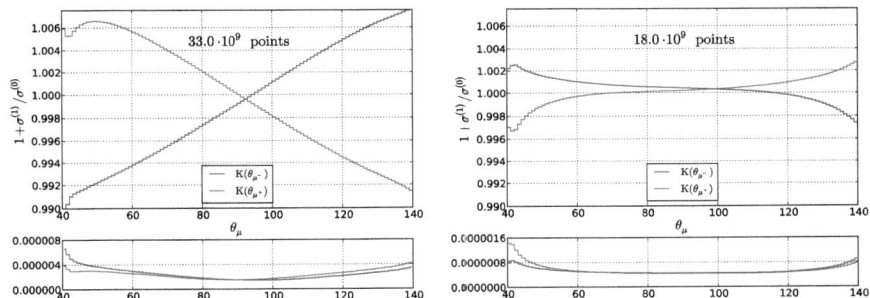

Figure 5.11.: "Penta" contribution to angular distributions of μ^+ and μ^- for BaBar setup (left σ, right $\bar{\sigma}$). Bottom: absolute error estimate.

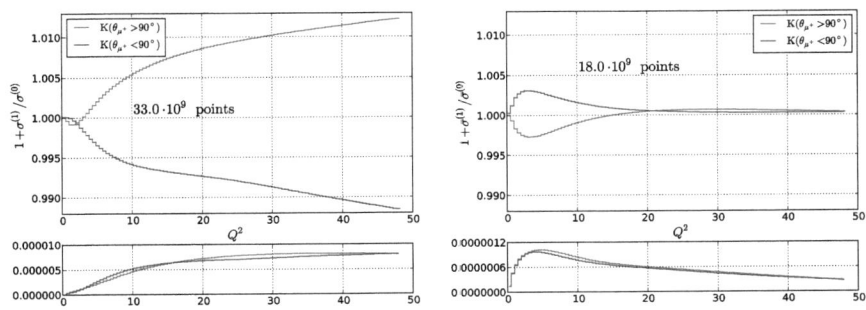

Figure 5.12.: "Penta" contribution to forward-backward asymmetry of μ^+ for BaBar setup (left σ, right $\bar{\sigma}$). Bottom: absolute error estimate.

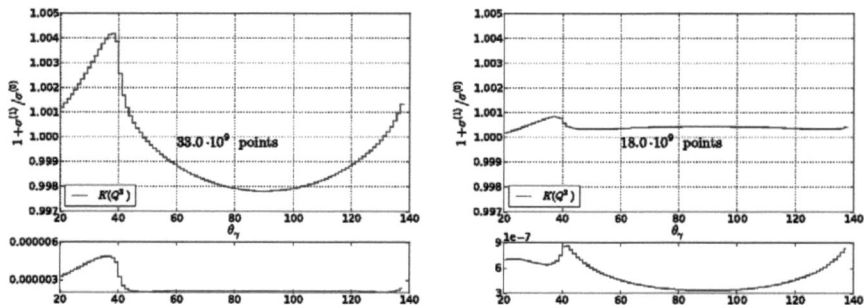

Figure 5.13.: "Penta" contribution to the angular distribution of the photon for BaBar setup (left σ, right $\bar{\sigma}$). Bottom: absolute error estimate.

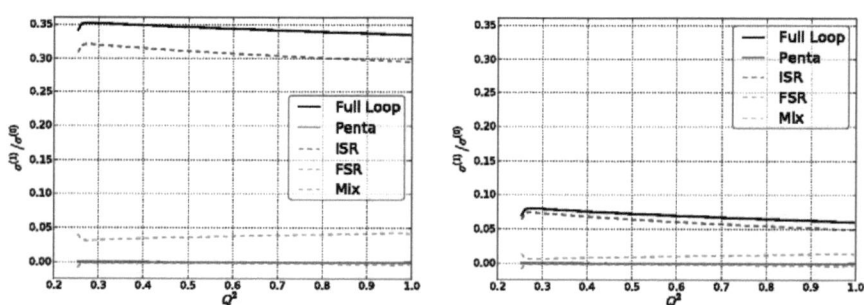

Figure 5.14.: Relative size of different contributions to the Q^2 distribution normalized on Born for KLOE setup (left σ, right $\bar{\sigma}$).

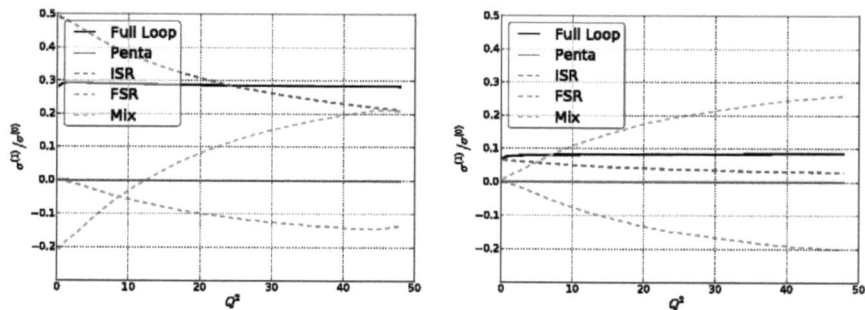

Figure 5.15.: Relative size of different contributions to the Q^2 distribution normalized on Born for BaBar setup (left σ, right $\bar{\sigma}$).

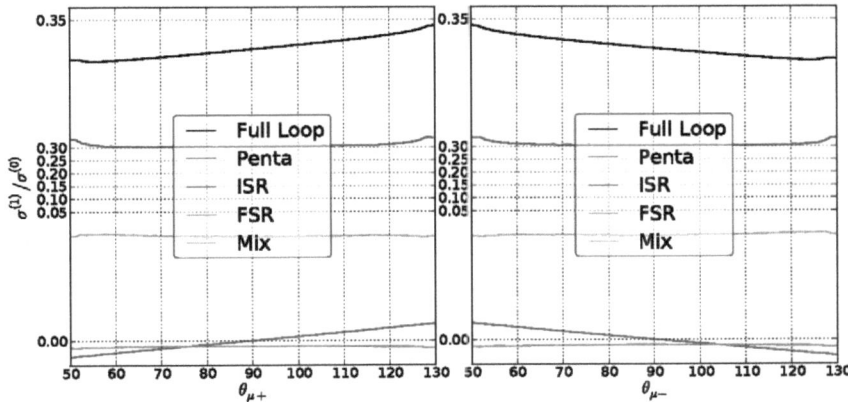

Figure 5.16.: Relative size of different contributions to angular distributions of μ^+ and μ^- normalized on Born for KLOE setup (for the plain σ (5.25)).

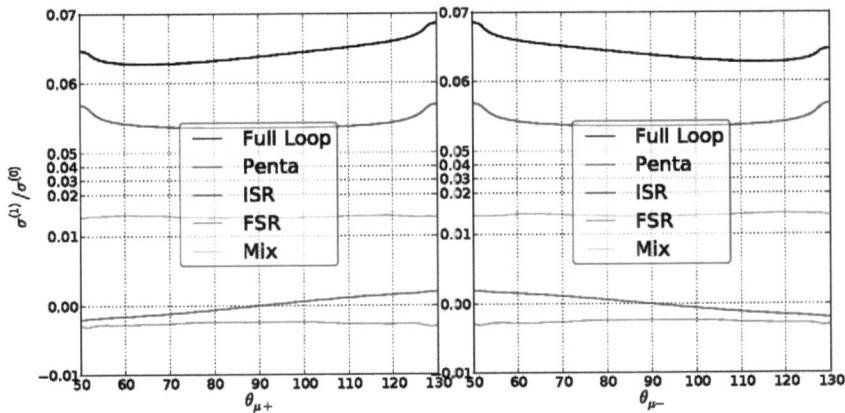

Figure 5.17.: Relative size of different contributions to angular distributions of μ^+ and μ^- normalized on Born for KLOE setup (for the subtracted $\bar{\sigma}$ (5.26)).

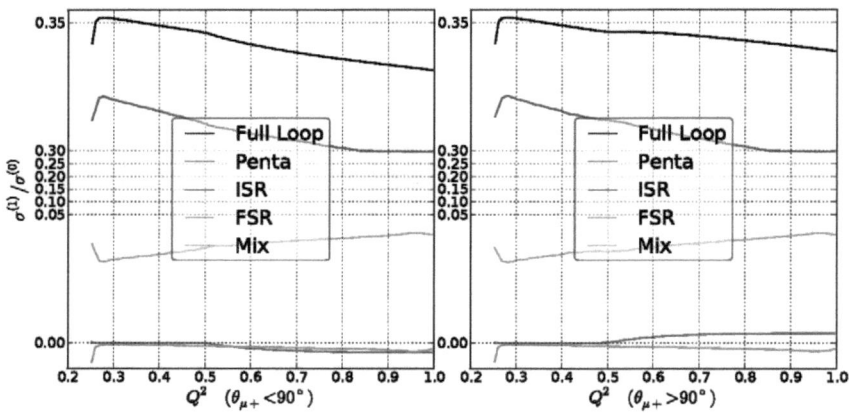

Figure 5.18.: Relative size of different contributions to forward-backward asymmetry of μ^+ normalized on Born for KLOE setup (for the plain σ (5.25)).

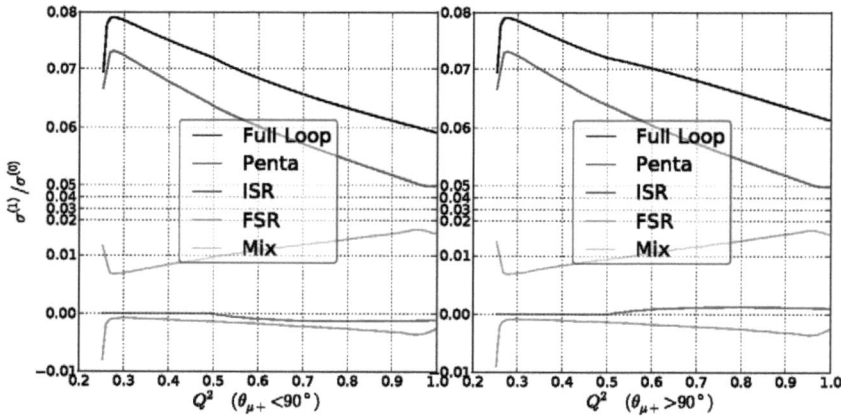

Figure 5.19.: Relative size of different contributions to forward-backward asymmetry of μ^+ normalized on Born for KLOE setup (for the subtracted $\bar{\sigma}$ (5.26)).

6. Summary and outlook

In this thesis we have discussed the problem of the stable evaluation of tensor integrals, which is a fundamental component of Feynman diagram-based one-loop computation.

We presented an open source computer library PJFry which implements the reduction scheme based on the method of integrals in shifted dimension. The library performs a numerical reduction of tensor pentagon integrals up to rank 5 to the set of basis scalar integrals for any physically relevant kinematic configuration. The reduction procedure completely avoids 5-point Gram determinants. The 4-point Gram determinants are dealt with by extending the integral basis with scalar integrals in shifted dimension. The evaluation of new basis integrals in the region of small Gram determinants is done by series expansion. The cache system ensures the efficient reuse of repeating building blocks transparently to the user. The automatic selection of the appropriate reduction formulae is done by means of a number of cutoff parameters, which have been tuned for optimized accuracy and performance in the series of tests.

The correctness of the implementation has been checked by several synthetic contraction tests for a selection of massless and massive kinematic configurations. We have demonstrated a proof-of-concept application to the evaluation of five gluon helicity amplitudes. They contain the highest rank tensor integrals among $2 \to 3$ amplitudes and thus are quite challenging for the Feynman diagrammatic approach. We have compared our toy calculation with the results of the generalized unitarity package NGluon [13] and found complete agreement.

Finally we applied PJFry to the calculation of one-loop QED virtual corrections to photon associated muon pair production at e^+e^- colliders. We have extensively studied the obtained expressions for numerical stability. The known ISR and FSR parts of the amplitude have been cross-checked with their implementation in the Monte-Carlo event generator Phokhara [66]. We presented virtual corrections to differential distributions for several observables commonly used in the settings of the detectors BaBar and KLOE. A separate set of plots have been produced for the new pieces containing pentagon diagrams and the corresponding box diagrams.

From the smooth shape of the distributions and steady decline of the statistical errors with increasing number of points we draw the conclusion that there are no significant numerical instabilities and that the code is ready for phenomenological applications.

The analysis of the plots shows that the muon angular dependence is the distribution with highest sensibility to the virtual pentagon corrections, especially in the case of KLOE. This gives a motivation for the inclusion of pentagon contributions into Monte-Carlo event generators together with the corresponding real corrections.

Outlook

The results of this thesis open a number of possibilities for future work. The tensor reduction program is ready for application to $2 \rightarrow 3$ calculations involving massless and massive particles. The tensor pentagons, being the most complicated objects at the one loop level, constitute a basis for higher point integrals. The extension to 6-point functions is not complicated and will cover the most of phenomenologically relevant processes. Due to the generic nature of the reduction formalism, the addition of complex masses is also straightforward. It will allow to compute a wide range of processes with unstable intermediate states.

The inclusion of pentagon contributions to $e^+e^- \rightarrow \mu^+\mu^-\gamma$ into Monte-Carlo event generators will allow to perform a complete NLO analysis of this process and to make a conclusive statement about the numerical significance of the pentagon part for various observables. This contribution seems to be on the edge of the current experimental precision and might become more relevant when more accurate measurements become available.

Appendices

A. Algebra of signed minors

A.1 Definition and general properties

Let M be a square $n \times n$ matrix. Let us specify r rows $i_1, \ldots, i_r$ and r columns $j_1, \ldots, j_r$. The intersection of these rows and columns forms an $r \times r$ square matrix. Such matrix is called a minor of order r and we denote it as

$$M^{i_1,i_2,\ldots,i_r}_{j_1,j_2,\ldots,j_r} \tag{A.1}$$

The $(n-r) \times (n-r)$ squared matrix with discarded r rows and columns is called complementary minor and denoted as:

$$M^{\{i_1,i_2,\ldots,i_r\}}_{\{j_1,j_2,\ldots,j_r\}} \tag{A.2}$$

The cofactor (or signed minor) of the minor of order r is the coefficient of Laplace theorem expansion for the determinant of M:

$$\det M = \sum_{\{C_{i_r}\} \text{ or } \{C_{j_r}\}} \det\left(M^{i_1,i_2,\ldots,i_r}_{j_1,j_2,\ldots,j_r}\right) \begin{pmatrix} i_1 & i_2 & \ldots & i_r \\ j_1 & j_2 & \ldots & j_r \end{pmatrix} \tag{A.3}$$

where the sum is taken over $\binom{n}{r}$ distinct r-combinations of either rows $i_1, \ldots, i_r$ or columns $j_1, \ldots, j_r$. The explicit expression for a signed minor is given in (3.5).

For completeness we define the minor of order 0 to be equal to 1. Then the cofactor of the minor of order 0 is just a determinant of the corresponding matrix:

$$M^0_0 \equiv 1 \qquad M^{\{\}}_{\{\}} \equiv M \qquad () = \det M$$

Signed minors are antisymmetric in any pair of adjacent row or column indices

$$\begin{pmatrix} i_1 \ldots i_k\, i_{k+1} \ldots i_r \\ j_1 \ldots j_k\, j_{k+1} \ldots j_r \end{pmatrix} = -\begin{pmatrix} i_1 \ldots i_{k+1}\, i_k \ldots i_r \\ j_1 \ldots j_k\, j_{k+1} \ldots j_r \end{pmatrix} = -\begin{pmatrix} i_1 \ldots i_k\, i_{k+1} \ldots i_r \\ j_1 \ldots j_{k+1}\, j_k \ldots j_r \end{pmatrix} \tag{A.4}$$

for any $1 \leq k \leq r-1$.

Signed minors of order r satisfy $r-1$ relations of the form

$$\det{}_{[\alpha\beta]} \begin{pmatrix} i_\alpha \\ j_\beta \end{pmatrix} = \begin{pmatrix} i_1 \ldots i_r \\ j_1 \ldots j_r \end{pmatrix} ()^{r-1} \tag{A.5}$$

where the term on the left hand side is the determinant of the matrix indexed by α and β. Which for $r=2$ gives

$$\begin{vmatrix} \begin{pmatrix} i_1 \\ j_1 \end{pmatrix} & \begin{pmatrix} i_1 \\ j_2 \end{pmatrix} \\ \begin{pmatrix} i_2 \\ j_1 \end{pmatrix} & \begin{pmatrix} i_2 \\ j_2 \end{pmatrix} \end{vmatrix} = \begin{pmatrix} i_1\, i_2 \\ j_1\, j_2 \end{pmatrix} () \tag{A.6}$$

The extension property of signed minors states that, any relation which is valid for () is also valid for any minor of ().

A.2 Symmetric bordered determinants

Since we use the algebra of signed minors mostly with the modified Cayley matrix, it is worth to write down additional special relations which are valid for this special case.

A bordered symmetric $(n+1) \times (n+1)$ matrix is defined as

$$Y_{ij} = \begin{cases} Y_{ij}, & i,j = 1,\ldots,n \\ 1, & i = 0, \; j = 1,\ldots,n \text{ or } j = 0, \; i = 1,\ldots,n \\ 0, & i,j = 0 \end{cases}$$

The signed minors are symmetric under switching row indices with column indices

$$\begin{pmatrix} i_1 i_2 \ldots i_r \\ j_1 j_2 \ldots j_r \end{pmatrix}_n = \begin{pmatrix} j_1 j_2 \ldots j_r \\ i_1 i_2 \ldots i_r \end{pmatrix}_n$$

The row expansion of $(\;)_n$ together with (A.6) gives

$$\sum_{i=1}^{n} \begin{pmatrix} i \\ 0 \end{pmatrix}_n = (\;)_n \tag{A.7}$$

$$\sum_{i=1}^{n} \begin{pmatrix} i \\ j \end{pmatrix}_n = 0, \qquad j \neq 0 \tag{A.8}$$

The master formula (3.6) is a consequence of the above together with (A.6) and the extension property:

$$\begin{pmatrix} \alpha \\ i \end{pmatrix}_n \begin{pmatrix} \alpha\,\beta \\ j\,0 \end{pmatrix}_n = \sum_{k=1}^{n} \begin{pmatrix} \alpha\,k \\ i\,0 \end{pmatrix}_n \begin{pmatrix} \alpha\,\beta \\ j\,0 \end{pmatrix}_n = \sum_{k=1}^{n} \begin{pmatrix} \alpha\,k \\ 0\,i \end{pmatrix}_n \begin{pmatrix} \alpha\,\beta \\ 0\,j \end{pmatrix}_n$$

$$= \sum_{k=1}^{n} \left[\begin{pmatrix} \alpha\,k \\ 0\,j \end{pmatrix}_n \begin{pmatrix} \alpha\,\beta \\ 0\,i \end{pmatrix}_n + \begin{pmatrix} \alpha\,k\,\beta \\ 0\,i\,j \end{pmatrix}_n \begin{pmatrix} \alpha \\ 0 \end{pmatrix}_n \right] = -\begin{pmatrix} \alpha \\ j \end{pmatrix}_n \begin{pmatrix} \alpha\,\beta \\ 0\,i \end{pmatrix}_n - \begin{pmatrix} \alpha \\ 0 \end{pmatrix}_n \begin{pmatrix} \alpha\,\beta \\ i\,j \end{pmatrix}_n$$

B. Scalar integral recurrence relations

In this section we list for completeness all relevant recurrence relations for the evaluation of boxes, triangles and bubbles.

For brevity in this section we omit combinatorial factors $n_{...}$ in front of the integrals

$$n_{ij} I^{[2l]}_{4,ij} \to \bar{I}^{[2l]}_{4,ij}, \qquad n_{ijk} I^{[2l]}_{4,ijk} \to \bar{I}^{[2l]}_{4,ijk}, \qquad n_{ijkl} I^{[2l]}_{4,ijkl} \to \bar{I}^{[2l]}_{4,ijkl}.$$

B.1 Boxes

In this section all signed minors come from 4-point kinematics and $()$ should be thought as $()_4$.

B.1.1 Downward recurrence

Explicit form of (3.23) and (3.24) for the box form-factor recursion. Each step reduces dimension and number of indices, but introduces an inverse 4-point Gram determinant $()$.

$$I^{[2(l+1)]}_{4} = \frac{1}{()}\frac{1}{d+2l-3}\left[\binom{0}{0}I^{[2l]}_{4} - \sum_{t=1}^{4}\binom{t}{0}I^{[2l],t}_{3}\right] \tag{B.1a}$$

$$I^{[2(l+1)]}_{4,i} = \frac{1}{()}\left[-\binom{0}{i}I^{[2l]}_{4} + \sum_{t=1}^{4}\binom{t}{i}I^{[2l],t}_{3}\right] \tag{B.1b}$$

$$\bar{I}^{[2(l+1)]}_{4,ij} = \frac{1}{()}\left[-\binom{0}{j}I^{[2l]}_{4,i} + \sum_{t=1}^{4}\binom{t}{j}I^{[2l],t}_{3,i} + \binom{i}{j}I^{[2l]}_{4}\right] \tag{B.1c}$$

$$\bar{I}^{[2(l+1)]}_{4,ijk} = \frac{1}{()}\left[-\binom{0}{k}\bar{I}^{[2l]}_{4,ij} + \sum_{t=1}^{4}\binom{t}{k}\bar{I}^{[2l],t}_{3,ij} + \binom{i}{k}I^{[2l]}_{4,j} + \binom{j}{k}I^{[2l]}_{4,i}\right] \tag{B.1d}$$

$$\bar{I}^{[2(l+1)]}_{4,ijkl} = \frac{1}{()}\left[-\binom{0}{l}\bar{I}^{[2l]}_{4,ijk} + \sum_{t=1}^{4}\binom{t}{l}\bar{I}^{[2l],t}_{3,ijk} + \binom{i}{l}\bar{I}^{[2l]}_{4,jk} + \binom{j}{l}\bar{I}^{[2l]}_{4,ik} + \binom{k}{l}\bar{I}^{[2l]}_{4,ij} \right]$$
(B.1e)

B.1.2 Index recurrence

Alternative recurrence scheme described in Sec. 3.3.3. The inverse Gram determinant is replaced by the Cayley determinant $\binom{0}{0}$, at cost of not reducing dimension and introducing a rational term.

$$I^{[2(l+1)]}_{4,i} = \frac{1}{\binom{0}{0}}\left[-(d+2l-3)\binom{0}{i}I^{[2(l+1)]}_{4} + \sum_{t=1}^{4}\binom{0t}{0i}I^{[2l],t}_{3} \right] \quad \text{(B.2a)}$$

$$\bar{I}^{[2(l+1)]}_{4,ij} = \frac{1}{\binom{0}{0}}\left[-(d+2l-4)\binom{0}{j}I^{[2(l+1)]}_{4,i} + \sum_{t=1}^{4}\binom{0t}{0j}I^{[2l],t}_{3,i} + \binom{0i}{0j}I^{[2l]}_{4} \right] \quad \text{(B.2b)}$$

$$\bar{I}^{[2(l+1)]}_{4,ijk} = \frac{1}{\binom{0}{0}}\left[-(d+2l-5)\binom{0}{k}\bar{I}^{[2(l+1)]}_{4,ij} + \sum_{t=1}^{4}\binom{0t}{0k}\bar{I}^{[2l],t}_{3,ij} \right.$$
$$\left. + \binom{0i}{0k}I^{[2l]}_{4,j} + \binom{0j}{0k}I^{[2l]}_{4,i} \right] \quad \text{(B.2c)}$$

$$\bar{I}^{[2(l+1)]}_{4,ijkl} = \frac{1}{\binom{0}{0}}\left[-(d+2l-6)\binom{0}{l}\bar{I}^{[2(l+1)]}_{4,ijk} + \sum_{t=1}^{4}\binom{0t}{0l}\bar{I}^{[2l],t}_{3,ijk} \right.$$
$$\left. + \binom{0i}{0l}\bar{I}^{[2l]}_{4,jk} + \binom{0j}{0l}\bar{I}^{[2l]}_{4,ik} + \binom{0k}{0l}\bar{I}^{[2l]}_{4,ij} \right] \quad \text{(B.2d)}$$

B.2 Triangles

In this section all signed minors come from 3-point kinematics and determinant () should be thought as ()$_3$.

B.2.1 Downward recurrence

Explicit form of (3.23) and (3.24) for the triangle form-factor recursion. Each step reduces dimension and number of indices, but introduces an inverse 3-point Gram determinant ().

$$I_3^{[2(l+1)]} = \frac{1}{()}\frac{1}{d+2l-2}\left[\begin{pmatrix}0\\0\end{pmatrix}I_3^{[2l]} - \sum_{u=1}^{3}\begin{pmatrix}u\\0\end{pmatrix}I_2^{[2l],u}\right] \tag{B.3a}$$

$$I_{3,i}^{[2(l+1)]} = \frac{1}{()}\left[-\begin{pmatrix}0\\i\end{pmatrix}I_3^{[2l]} + \sum_{u=1}^{3}\begin{pmatrix}u\\i\end{pmatrix}I_2^{[2l],u}\right] \tag{B.3b}$$

$$\bar{I}_{3,ij}^{[2(l+1)]} = \frac{1}{()}\left[-\begin{pmatrix}0\\j\end{pmatrix}I_{3,i}^{[2l]} + \sum_{u=1}^{3}\begin{pmatrix}u\\j\end{pmatrix}I_{2,i}^{[2l],u} + \begin{pmatrix}i\\j\end{pmatrix}I_3^{[2l]}\right] \tag{B.3c}$$

$$\bar{I}_{3,ijk}^{[2(l+1)]} = \frac{1}{()}\left[-\begin{pmatrix}0\\k\end{pmatrix}\bar{I}_{3,ij}^{[2l]} + \sum_{u=1}^{3}\begin{pmatrix}u\\k\end{pmatrix}\bar{I}_{2,ij}^{[2l],u} + \begin{pmatrix}i\\k\end{pmatrix}I_{3,j}^{[2l]} + \begin{pmatrix}j\\k\end{pmatrix}I_{3,i}^{[2l]}\right] \tag{B.3d}$$

B.2.2 Index recurrence

Alternative recurrence scheme described in Sec. 3.3.3. The inverse Gram determinant is replaced by the Cayley determinant $\begin{pmatrix}0\\0\end{pmatrix}$, at cost of not reducing dimension and introducing a rational term.

Valid only for IR finite triangles $\begin{pmatrix}0\\0\end{pmatrix} \neq 0$.

$$I_{3,i}^{[2(l+1)]} = \frac{1}{\begin{pmatrix}0\\0\end{pmatrix}}\left[-(d+2l-2)\begin{pmatrix}0\\i\end{pmatrix}I_3^{[2(l+1)]} + \sum_{u=1}^{3}\begin{pmatrix}0u\\0i\end{pmatrix}I_2^{[2l],u}\right] \tag{B.4a}$$

$$\bar{I}_{3,ij}^{[2(l+1)]} = \frac{1}{\begin{pmatrix}0\\0\end{pmatrix}}\left[-(d+2l-3)\begin{pmatrix}0\\j\end{pmatrix}I_{3,i}^{[2(l+1)]} + \sum_{u=1}^{3}\begin{pmatrix}0u\\0j\end{pmatrix}I_{2,i}^{[2l],u} + \begin{pmatrix}0i\\0j\end{pmatrix}I_3^{[2l]}\right] \tag{B.4b}$$

$$\bar{I}_{3,ijk}^{[2(l+1)]} = \frac{1}{\begin{pmatrix}0\\0\end{pmatrix}}\left[-(d+2l-4)\begin{pmatrix}0\\k\end{pmatrix}\bar{I}_{3,ij}^{[2(l+1)]} + \sum_{u=1}^{3}\begin{pmatrix}0u\\0k\end{pmatrix}I_{2,ij}^{[2l],u}\right.$$
$$\left. + \begin{pmatrix}0i\\0k\end{pmatrix}I_{3,j}^{[2l]} + \begin{pmatrix}0j\\0k\end{pmatrix}I_{3,i}^{[2l]}\right] \tag{B.4c}$$

B.2.3 IR case recurrence

Recurrence scheme for IR divergent triangles where $\binom{0}{0} = 0$. Reduces dimension and number of legs, but not the number of indices. Requires bubbles in 2 dimensions.

Here the last index is free. Its value should be chosen such that the denominator is not small or zero $\binom{0}{k} \neq 0$.

$$I_3^{[2(l+1)]} = \frac{1}{\binom{0}{i}} \frac{1}{(d+2l-2)} \left[\sum_{u=1}^{3} \binom{0u}{0i} I_2^{[2l],u} \right] \tag{B.5a}$$

$$I_{3,i}^{[2(l+1)]} = \frac{1}{\binom{0}{j}} \frac{1}{(d+2l-3)} \left[\sum_{u=1}^{3} \binom{0u}{0j} I_{2,i}^{[2l],u} + \binom{0i}{0j} I_3^{[2l]} \right] \tag{B.5b}$$

$$\bar{I}_{3,ij}^{[2(l+1)]} = \frac{1}{\binom{0}{k}} \frac{1}{(d+2l-4)} \left[\sum_{u=1}^{3} \binom{0u}{0k} \bar{I}_{2,ij}^{[2l],u} + \binom{0i}{0k} I_{3,j}^{[2l]} + \binom{0j}{0k} I_{3,i}^{[2l]} \right] \tag{B.5c}$$

$$\bar{I}_{3,ijk}^{[2(l+1)]} = \frac{1}{\binom{0}{l}} \frac{1}{(d+2l-5)} \left[\sum_{u=1}^{3} \binom{0u}{0l} \bar{I}_{2,ijk}^{[2l],u} + \binom{0i}{0l} \bar{I}_{3,jk}^{[2l]} + \binom{0j}{0l} \bar{I}_{3,ik}^{[2l]} + \binom{0k}{0l} \bar{I}_{3,ij}^{[2l]} \right] \tag{B.5d}$$

B.3 Bubbles

$$I_2^{[2(l+1)]} = \frac{1}{()} \frac{1}{d+2l-3} \left[\binom{0}{0} I_2^{[2l]} - \sum_{v=1}^{2} \binom{v}{0} I_1^{[2l],v} \right] \tag{B.6a}$$

$$I_{2,i}^{[2(l+1)]} = \frac{1}{()} \left[-\binom{0}{i} I_2^{[2l]} + \sum_{v=1}^{2} \binom{u}{i} I_1^{[2l],v} \right] \tag{B.6b}$$

$$\bar{I}_{2,ij}^{[2(l+1)]} = \frac{1}{()} \left[-\binom{0}{j} I_{2,i}^{[2l]} + \sum_{v=1}^{2} \binom{u}{j} I_{1,i}^{[2l],v} + \binom{i}{j} I_2^{[2l]} \right] \tag{B.6c}$$

See App. C for the tadpoles.

C. On-shell two point functions

In this appendix we use the notation of Chapter 4.

All order expressions for 1- and 2-point functions in generic dimension $d = 4 - 2\epsilon$

$$I^{[-2\epsilon]}_{1,(n_1)}(m) = \frac{(-1)^{n_1}(m)^{2-\epsilon-n_1}\Gamma(\epsilon-2+n_1)}{\Gamma(n_1)} \tag{C.1}$$

$$I^{[-2\epsilon]}_{2,(n_1,n_2)}(m,m,0) = \frac{(-1)^{n_1+n_2}(m)^{2-\epsilon-n_1-n_2}\Gamma(4-2\epsilon-n_1-2n_2)\Gamma(\epsilon-2+n_1+n_2)}{\Gamma(n_1)\Gamma(4-2\epsilon-n_1-n_2)}$$

$$I^{[-2\epsilon]}_{2,(n_1,n_2)}(0,m,m) = \frac{(-1)^{n_1+n_2}(m)^{2-\epsilon-n_1-n_2}\Gamma(\epsilon-2+n_1+n_2)}{\Gamma(n_1+n_2)}$$

$$I^{[-2\epsilon]}_{2,(1,1)}(0,m_1,m_2) = -\frac{(m_1^{1-\epsilon}-m_2^{1-\epsilon})\Gamma(\epsilon-1)}{m_1^2-m_2^2}$$

$$I^{[-2\epsilon]}_{2,(1,2)}(0,m_1,m_2) = \frac{m_1^{-\epsilon}m_2^{-\epsilon}(m_1^\epsilon(m_1+\epsilon(m_2-m_1))+m_1 m_2^\epsilon)\Gamma(1-\epsilon)\Gamma(\epsilon)}{(m_1-m_2)^2\Gamma(2-\epsilon)}$$

$$I^{[-2\epsilon]}_{2,(2,1)}(0,m_1,m_2) = \frac{m_1^{-\epsilon}m_2^{-\epsilon}(m_2 m_1^\epsilon - m_2^\epsilon(\epsilon(m_1-m_2)+m_2))\Gamma(1-\epsilon)\Gamma(\epsilon)}{(m_1-m_2)^2\Gamma(2-\epsilon)}$$

$$I^{[-2\epsilon]}_{2,(2,2)}(0,m_1,m_2) = \frac{(m_2^\epsilon((\epsilon-1)m_1-(\epsilon+1)m_2)+m_1^\epsilon((\epsilon+1)m_1-(\epsilon-1)m_2))\Gamma(\epsilon)}{(m_1-m_2)^3(\epsilon-1)}$$

C.1 Notable relations

$$I^{[2]}_1(m) = -\frac{2m}{d}I_1(m)$$

$$I_2(m,m,0) = (m)^{-\epsilon}\Gamma(\epsilon)/(1-2\epsilon) = m^{-1}(1-\epsilon)I_1(m) + O(\epsilon)$$

$$I^{[2]}_2(m,m,0) = (m)^{1-\epsilon}\Gamma(\epsilon-1)/(3-2\epsilon) = -I_1(m)/(d-1) = -\frac{1}{9}(3+2\epsilon)I_1(m) + O(\epsilon)$$

$$I_2(0,m,m) = (m)^{-\epsilon}\Gamma(\epsilon) = m^{-1}(1-\epsilon)I_1(m)$$

$$I^{[2]}_2(0,m,m) = (m)^{1-\epsilon}\Gamma(\epsilon-1) = (-2mI_2(0,m,m) - I_1(m))/(d-1) = -I_1(m)$$

$$I^{[2]}_{2,i}(0,m,m) = -\frac{1}{2}(m)^{-\epsilon}\Gamma(\epsilon) = -\frac{1}{2}I_2(0,m,m)$$

$$I_{2,i}^{[4]}(0,m,m) = -\frac{1}{2}(m)^{1-\epsilon}\Gamma(\epsilon-1) = \frac{1}{2}I_1(m)$$
$$I_{2,ij}^{[4]}(0,m,m) = \frac{1}{6}(m)^{-\epsilon}\Gamma(\epsilon) = -\frac{1}{3}I_{2,i}^{[2]}(0,m,m) = \frac{1}{6}I_2(0,m,m)$$

C.2 Special cases

Bubble in $4 + 2l - 2\epsilon$ dimensions with zero momentum transfer and different masses, I_2 with $s = 0$, $m_1 \neq m_2$:

$$I_2^{[-2]}(0,m_1,m_2) = -\frac{m_1^{-1}I_1(m_1) - m_2^{-1}I_1(m_2)}{(m_1 - m_2)} + O(\epsilon)$$

$$I_2^{[0]}(0,m_1,m_2) = +\frac{I_1(m_1) - I_1(m_2)}{m_1 - m_2} + O(\epsilon)$$

$$I_2^{[2]}(0,m_1,m_2) = -\frac{1}{4}(m_1 + m_2) - \frac{m_1 I_1(m_1) - m_2 I_1(m_2)}{2(m_1 - m_2)} + O(\epsilon)$$

$$I_2^{[4]}(0,m_1,m_2) = +\frac{5}{36}(m_1^2 + m_1 m_2 + m_2^2) + \frac{m_1^2 I_1(m_1) - m_2^2 I_1(m_2)}{6(m_1 - m_2)} + O(\epsilon)$$

$$I_2^{[6]}(0,m_1,m_2) = -\frac{13}{288}(m_1 + m_2)(m_1^2 + m_2^2) - \frac{m_1^3 I_1(m_1) - m_2^3 I_1(m_2)}{24(m_1 - m_2)} + O(\epsilon)$$

$$I_2^{[8]}(0,m_1,m_2) = +\frac{77(m_1^5 - m_2^5)}{7200(m_1 - m_2)} + \frac{m_1^4 I_1(m_1) - m_2^4 I_1(m_2)}{120(m_1 - m_2)} + O(\epsilon)$$

$$I_2^{[10]}(0,m_1,m_2) = -\frac{29(m_1^6 - m_2^6)}{120^2(m_1 - m_2)} - \frac{m_1^5 I_1(m_1) - m_2^5 I_1(m_2)}{6!(m_1 - m_2)} + O(\epsilon)$$

$$I_2^{[12]}(0,m_1,m_2) = +\frac{223(m_1^7 - m_2^7)}{840^2(m_1 - m_2)} + \frac{m_1^6 I_1(m_1) - m_2^6 I_1(m_2)}{7!(m_1 - m_2)} + O(\epsilon)$$

$$I_2^{[14]}(0,m_1,m_2) = -\frac{481(m_1^8 - m_2^8)}{3360^2(m_1 - m_2)} - \frac{m_1^8 I_1(m_1) - m_2^8 I_1(m_2)}{8!(m_1 - m_2)} + O(\epsilon)$$

Bubble in $4 + 2l - 2\epsilon$ dimensions with zero momentum transfer and equal masses, I_2 with $s = 0$, $m_1 = m_2 = m$:

$$I_2^{[-2]}(0,m,m) = m^{-1}$$
$$I_2^{[0]}(0,m,m) = -(\epsilon - 1)m^{-1}I_1(m) = -1 + m^{-1}I_1(m) + O(\epsilon)$$
$$I_2^{[2]}(0,m,m) = -I_1(m)$$
$$I_2^{[4]}(0,m,m) = -(\epsilon - 2)^{-1}mI_1(m) = \frac{1}{4}m(m + 2I_1(m)) + O(\epsilon)$$

$$I_2^{[6]}(0,m,m) = -(\epsilon-2)^{-1}(\epsilon-3)^{-1}m^2 I_1(m) = -\frac{1}{36}m^2(5m+6I_1(m)) + O(\epsilon)$$

$$\vdots$$

$$I_2^{[2l]}(0,m,m) = -m^{l-1}(\epsilon-1)(\epsilon)_{-l} I_1(m)$$

where $(\epsilon)_{-l}$ is the Pochhammer symbol.

Bubble in $4+2l-2\epsilon$ dimensions with zero momentum transfer and one zero mass, I_2 with $s=0$, $m_1=0$, $m_2=m$:

$$I_2^{[0]}(0,0,m) = m^{-1} I_1(m)$$

$$I_2^{[2]}(0,0,m) = (\epsilon-2)^{-1} I_1(m) = -\frac{1}{4}(m+2I_1(m)) + O(\epsilon)$$

$$I_2^{[4]}(0,0,m) = (\epsilon-2)^{-1}(\epsilon-3)^{-1} m I_1(m) = \frac{1}{36}m(5m+6I_1(m)) + O(\epsilon)$$

$$I_2^{[6]}(0,0,m) = (\epsilon-2)^{-1}(\epsilon-3)^{-1}(\epsilon-4)^{-1} m^2 I_1(m) = -\frac{1}{288}m^2(13m+12I_1(m)) + O(\epsilon)$$

$$\vdots$$

$$I_2^{[2l]}(0,0,m) = m^{l-1}(\epsilon-1)(\epsilon)_{-(l+1)} I_1(m)$$

where $(\epsilon)_{-(l+1)}$ is the Pochhammer symbol.

Bubble in $4+2l-2\epsilon$ dimensions with raised power on one propagator, zero momentum transfer and different masses, $I_{2,i}$ with $s=0$, $m_1 \neq m_2$:

$$I_{2,1}^{[0]}(0,m_1,m_2) = -\frac{1}{m_1-m_2} - \frac{m_2 m_1^{-1} I_1(m_1) - I_1(m_2)}{(m_1-m_2)^2} + O(\epsilon)$$

$$I_{2,2}^{[0]}(0,m_1,m_2) = \frac{1}{m_1-m_2} + \frac{I_1(m_1) - m_1 m_2^{-1} I_1(m_2)}{(m_1-m_2)^2} + O(\epsilon)$$

$$I_{2,1}^{[2]}(0,m_1,m_2) = \frac{m_1+m_2}{4(m_1-m_2)} - \frac{(m_1-2m_2)I_1(m_1) + m_2 I_1(m_2)}{2(m_1-m_2)^2} + O(\epsilon)$$

$$I_{2,2}^{[2]}(0,m_1,m_2) = -\frac{m_1+m_2}{4(m_1-m_2)} - \frac{m_1 I_1(m_1) + (m_2-2m_1)I_1(m_2)}{2(m_1-m_2)^2} + O(\epsilon)$$

$$I_{2,1}^{[4]}(0,m_1,m_2) = \frac{4m_1^2 - 5m_1 m_2 - 5m_2^2}{36(m_1-m_2)} + \frac{m_1(2m_1-3m_2)I_1(m_1) + m_2^2 I_1(m_2)}{6(m_1-m_2)^2} + O(\epsilon)$$

$$I_{2,2}^{[4]}(0,m_1,m_2) = -\frac{4m_2^2 - 5m_1 m_2 - 5m_1^2}{36(m_1-m_2)} + \frac{m_1^2 I_1(m_1) + m_2(2m_2-3m_1)I_1(m_2)}{6(m_1-m_2)^2} + O(\epsilon)$$

D. Diagrams for $e^+e^- \to \mu^+\mu^-\gamma$

The labeling is the following: [AnBC]

A loop by Initial, Final state or mixed with Same, Opposite direction

n number of loop legs

B radiation from Initial, Final, Loop line

C radiation off Particle, Anti-particle

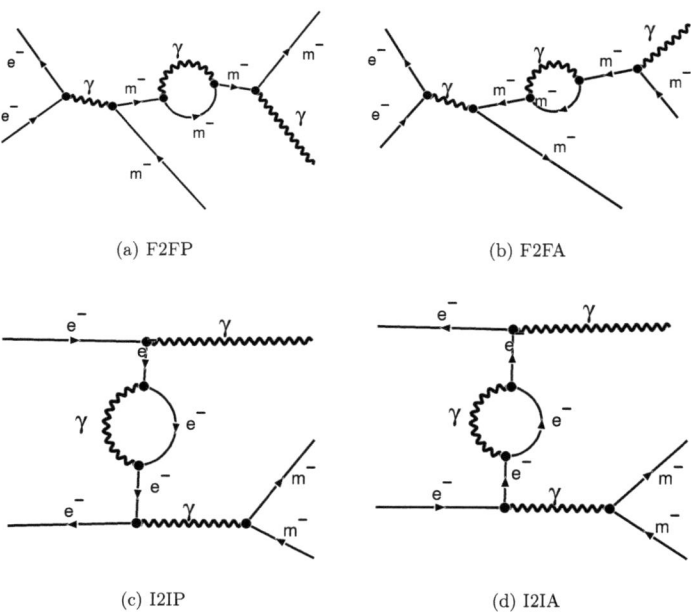

Figure D.1.: Bubble topology diagrams for $e^+e^- \to \mu^+\mu^-\gamma$

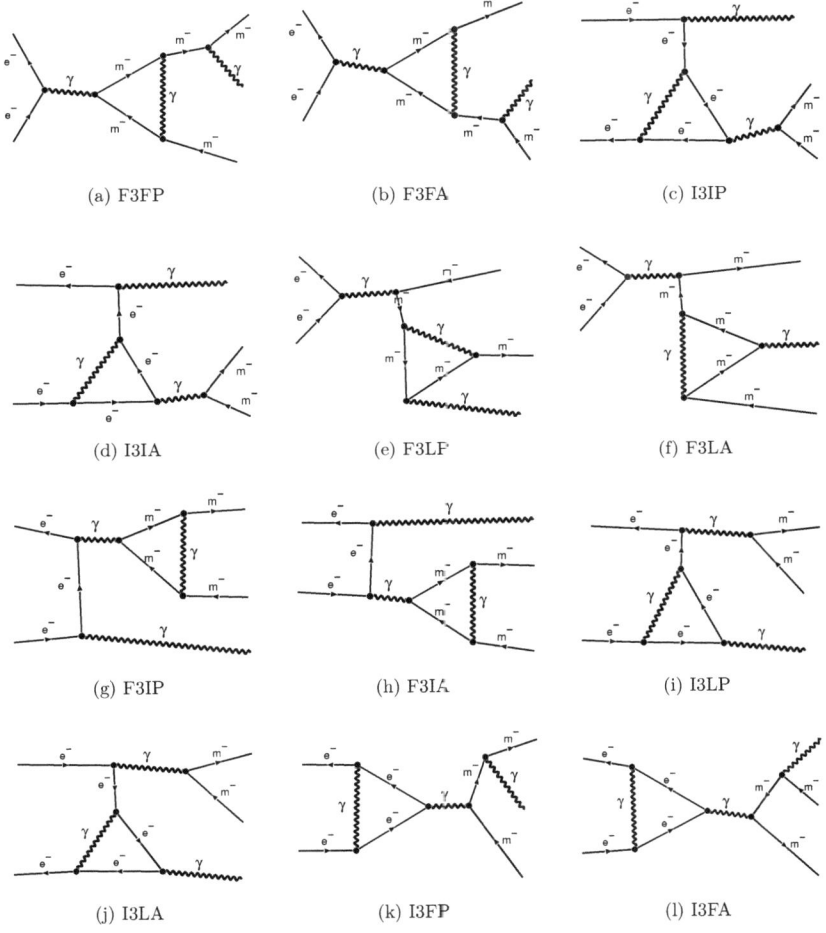

Figure D.2.: Triangle topology diagrams for $e^+e^- \to \mu^+\mu^-\gamma$

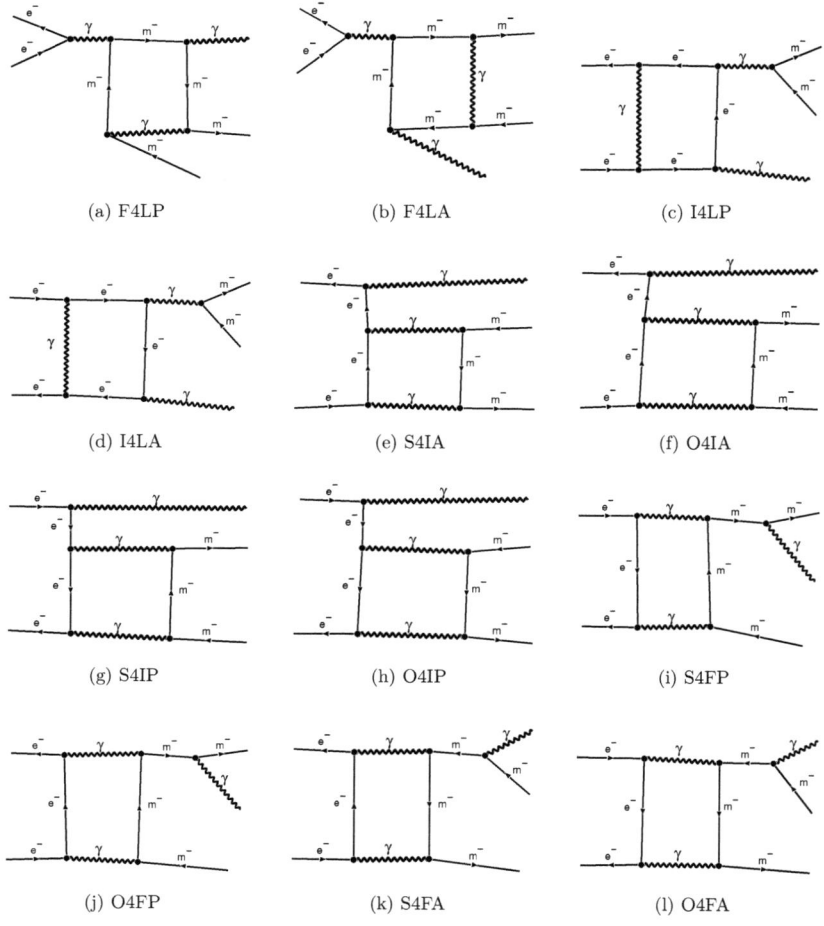

Figure D.3.: Box topology diagrams for $e^+e^- \to \mu^+\mu^-\gamma$

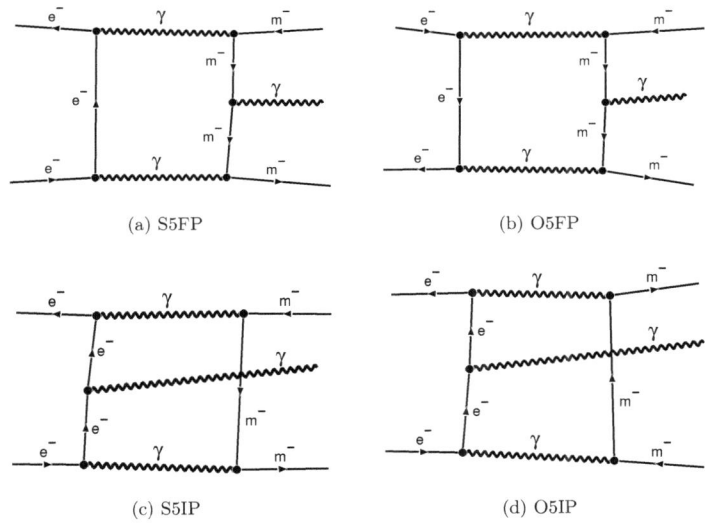

Figure D.4.: Pentagon topology diagrams for $e^+e^- \to \mu^+\mu^-\gamma$

Bibliography

[1] S. Actis et al. Quest for precision in hadronic cross sections at low energy: Monte Carlo tools vs. experimental data. *Eur. Phys. J.*, C66: 585–686, 2010. doi: 10.1140/epjc/s10052-010-1251-4.

[2] Stefano Actis, Pierpaolo Mastrolia, and Giovanni Ossola. Numerical evaluation of one-loop scattering amplitudes for hard-photon emission in Bhabha scattering. *Acta Phys. Polon.*, B40:2957–2963, 2009.

[3] Stefano Actis, Pierpaolo Mastrolia, and Giovanni Ossola. NLO QED Corrections to Hard-Bremsstrahlung Emission in Bhabha Scattering. *Phys. Lett.*, B682:419–427, 2010. doi: 10.1016/j.physletb.2009.11.035.

[4] Johan Alwall, Michel Herquet, Fabio Maltoni, Olivier Mattelaer, and Tim Stelzer. MadGraph 5 : Going Beyond. *JHEP*, 06:128, 2011. doi: 10.1007/JHEP06(2011)128.

[5] G. Amelino-Camelia et al. Physics with the KLOE-2 experiment at the upgraded DAΦNE. *Eur. Phys. J.*, C68:619–681, 2010. doi: 10.1140/epjc/s10052-010-1351-1.

[6] J. R. Andersen et al. The SM and NLO multileg working group: Summary report. arXiv:1003.1241 [hep-ph], 2010.

[7] A. B. Arbuzov and E. S. Scherbakova. QED collinear radiation factors in the next-to-leading logarithmic approximation. *Phys. Lett.*, B660: 37–42, 2008. doi: 10.1016/j.physletb.2007.12.038.

[8] A. B. Arbuzov, V. V. Bytev, and E. A. Kuraev. Radiative muon pair production in high energy electron positron annihilation process. *JETP Lett.*, 79:593–597, 2004. doi: 10.1134/1.1790013.

[9] A. B. Arbuzov, V. V. Bytev, E. A. Kuraev, E. Tomasi-Gustafsson, and Yu. M. Bystritskiy. Structure function approach in QED for high energy processes. *Phys. Part. Nucl.*, 41:394–424, 2010. doi: 10.1134/S1063779610030020.

[10] A. B. Arbuzov, V. V. Bytev, E. A. Kuraev, E. Tomasi-Gustafsson, and Yu. M. Bystritskiy. Exact results in QED. *Phys. Part. Nucl.*, 42:1–54, 2011. doi: 10.1134/S1063779611010023.

[11] A. B. Arbuzov et al. Large angle QED processes at e^+ e^- colliders at energies below 3 GeV. *JHEP*, 10:001, 1997.

[12] J. F. Ashmore. A Method of Gauge Invariant Regularization. *Lett. Nuovo Cim.*, 4:289–290, 1972.

[13] Simon Badger, Benedikt Biedermann, and Peter Uwer. NGluon: A Package to Calculate One-loop Multi-gluon Amplitudes. *Comput. Phys. Commun.*, 182:1674–1692, 2011. doi: 10.1016/j.cpc.2011.04.008.

[14] Simon Badger, Ralf Sattler, and Valery Yundin. One-Loop Helicity Amplitudes for ttbar Production at Hadron Colliders. *Phys. Rev.*, D83: 074020, 2011. doi: 10.1103/PhysRevD.83.074020.

[15] V. N. Baier and Valery A. Khoze. Photon emission in muon pair production in electron-positron collisions. *Sov. Phys. JETP*, 21:629–632, 1965.

[16] G. Belanger et al. Automatic calculations in high energy physics and Grace at one-loop. *Phys. Rept.*, 430:117–209, 2006. doi: 10.1016/j.physrep.2006.02.001.

[17] G. W. Bennett et al. Final report of the muon E821 anomalous magnetic moment measurement at BNL. *Phys. Rev.*, D73:072003, 2006. doi: 10.1103/PhysRevD.73.072003.

[18] Frits A. Berends and R. Kleiss. Distributions in the Process e^+ $e^- \to \mu^+$ μ^- (gamma). *Nucl. Phys.*, B177:237, 1981. doi: 10.1016/0550-3213(81)90390-4.

[19] Frits A. Berends, K. J. F. Gaemer, and R. Gastmans. Hard photon corrections for the process e^+ $e^- \to \mu^\pm$ $\mu^\mp$. *Nucl. Phys.*, B57:381–400, 1973. doi: 10.1016/0550-3213(73)90108-9.

[20] Frits A. Berends, G. J. H. Burgers, C. Mana, M. Martinez, and W. L. van Neerven. RADIATIVE CORRECTIONS TO THE PROCESS e^+ $e^- \to$ neutrino anti-neutrino gamma. *Nucl. Phys.*, B301:583, 1988. doi: 10.1016/0550-3213(88)90278-7.

[21] Frits A. Berends et al. MULTIPLE BREMSSTRAHLUNG IN GAUGE THEORIES AT HIGH-ENERGIES. 5. THE PROCESS $e^+ e^- \to \mu^+ \mu^-$ gamma gamma. *Nucl. Phys.*, B264:243, 1986. doi: 10.1016/0550-3213(86)90481-5.

[22] C. F. Berger et al. An Automated Implementation of On-Shell Methods for One- Loop Amplitudes. *Phys. Rev.*, D78:036003, 2008. doi: 10.1103/PhysRevD.78.036003.

[23] C. F. Berger et al. Precise Predictions for W + 4 Jet Production at the Large Hadron Collider. *Phys. Rev. Lett.*, 106:092001, 2011. doi: 10.1103/PhysRevLett.106.092001.

[24] Z. Bern and A. G. Morgan. Massive Loop Amplitudes from Unitarity. *Nucl. Phys.*, B467:479–509, 1996. doi: 10.1016/0550-3213(96)00078-8.

[25] Z. Bern et al. The NLO multileg working group: Summary report. arXiv:0803.0494 [hep-ph], 2008.

[26] Zvi Bern and David A. Kosower. The Computation of loop amplitudes in gauge theories. *Nucl. Phys.*, B379:451–561, 1992. doi: 10.1016/0550-3213(92)90134-W.

[27] Zvi Bern, Lance J. Dixon, and David A. Kosower. Dimensionally Regulated One-Loop Integrals. *Phys. Lett.*, B302:299–308, 1993. doi: 10.1016/0370-2693(93)90400-C.

[28] Zvi Bern, Lance J. Dixon, and David A. Kosower. One loop corrections to five gluon amplitudes. *Phys.Rev.Lett.*, 70:2677–2680, 1993. doi: 10.1103/PhysRevLett.70.2677.

[29] Zvi Bern, Lance J. Dixon, David C. Dunbar, and David A. Kosower. One-Loop n-Point Gauge Theory Amplitudes, Unitarity and Collinear Limits. *Nucl. Phys.*, B425:217–260, 1994. doi: 10.1016/0550-3213(94)90179-1.

[30] Zvi Bern, Lance J. Dixon, and David A. Kosower. Dimensionally regulated pentagon integrals. *Nucl. Phys.*, B412:751–816, 1994. doi: 10.1016/0550-3213(94)90398-0.

[31] Zvi Bern, Lance J. Dixon, David C. Dunbar, and David A. Kosower. Fusing gauge theory tree amplitudes into loop amplitudes. *Nucl. Phys.*, B435:59–101, 1995. doi: 10.1016/0550-3213(94)00488-Z.

[32] Zvi Bern, Lance J. Dixon, and David A. Kosower. Progress in one-loop QCD computations. *Ann. Rev. Nucl. Part. Sci.*, 46:109–148, 1996. doi: 10.1146/annurev.nucl.46.1.109.

[33] G. Bevilacqua, M. Czakon, C. G. Papadopoulos, R. Pittau, and M. Worek. Assault on the NLO Wishlist: $pp \to t\bar{t}b\bar{b}$. *JHEP*, 09:109, 2009. doi: 10.1088/1126-6708/2009/09/109.

[34] G. Bevilacqua, M. Czakon, C. G. Papadopoulos, and M. Worek. Dominant QCD Backgrounds in Higgs Boson Analyses at the LHC: A Study of $pp \to t\bar{t} + 2$ jets at Next-To-Leading Order. *Phys. Rev. Lett.*, 104: 162002, 2010. doi: 10.1103/PhysRevLett.104.162002.

[35] G. Bevilacqua et al. NLO QCD calculations with HELAC-NLO. *Nucl. Phys. Proc. Suppl.*, 205-206:211–217, 2010. doi: 10.1016/j.nuclphysbps.2010.08.045.

[36] Giuseppe Bevilacqua, Michal Czakon, Andreas van Hameren, Costas G. Papadopoulos, and Malgorzata Worek. Complete off-shell effects in top quark pair hadroproduction with leptonic decay at next-to-leading order. *JHEP*, 02:083, 2011. doi: 10.1007/JHEP02(2011)083.

[37] T. Binoth, J. P. Guillet, and G. Heinrich. Reduction formalism for dimensionally regulated one-loop N-point integrals. *Nucl. Phys.*, B572: 361–386, 2000. doi: 10.1016/S0550-3213(00)00040-7.

[38] T. Binoth, J. Ph. Guillet, G. Heinrich, E. Pilon, and C. Schubert. An algebraic / numerical formalism for one-loop multi-leg amplitudes. *JHEP*, 10:015, 2005. doi: 10.1088/1126-6708/2005/10/015.

[39] T. Binoth, J. Ph. Guillet, and G. Heinrich. Algebraic evaluation of rational polynomials in one-loop amplitudes. *JHEP*, 02:013, 2007. doi: 10.1088/1126-6708/2007/02/013.

[40] T. Binoth, G. Ossola, C. G. Papadopoulos, and R. Pittau. NLO QCD corrections to tri-boson production. *JHEP*, 06:082, 2008. doi: 10.1088/1126-6708/2008/06/082.

[41] T. Binoth, J.-Ph. Guillet, G. Heinrich, E. Pilon, and T. Reiter. Golem95: a numerical program to calculate one-loop tensor integrals with up to six external legs. *Comput. Phys. Commun.*, 180:2317–2330, 2009. doi: 10.1016/j.cpc.2009.06.024.

[42] T. Binoth et al. Next-to-leading order QCD corrections to $pp \to b\bar{b}b\bar{b}+X$ at the LHC: the quark induced case. *Phys. Lett.*, B685:293–296, 2010. doi: 10.1016/j.physletb.2010.02.010.

[43] F. Bloch and A. Nordsieck. Note on the Radiation Field of the electron. *Phys. Rev.*, 52:54–59, 1937. doi: 10.1103/PhysRev.52.54.

[44] M. Bohm, H. Spiesberger, and W. Hollik. On the One Loop Renormalization of the Electroweak Standard Model and Its Application to Leptonic Processes. *Fortsch. Phys.*, 34:687–751, 1986. doi: 10.1002/prop.19860341102.

[45] C. G. Bollini and J. J. Giambiagi. Lowest order divergent graphs in nu-dimensional space. *Phys. Lett.*, B40:566–568, 1972. doi: 10.1016/0370-2693(72)90483-2.

[46] E. Boos et al. CompHEP 4.4: Automatic computations from Lagrangians to events. *Nucl. Instrum. Meth.*, A534:250–259, 2004. doi: 10.1016/j.nima.2004.07.096.

[47] Fawzi Boudjema and Le Duc Ninh. b anti-b Higgs production at the LHC: Yukawa corrections and the leading Landau singularity. *Phys. Rev.*, D78:093005, 2008. doi: 10.1103/PhysRevD.78.093005.

[48] Axel Bredenstein, Ansgar Denner, Stefan Dittmaier, and Stefano Pozzorini. NLO QCD corrections to top anti-top bottom anti-bottom production at the LHC: 1. quark-antiquark annihilation. *JHEP*, 08:108, 2008. doi: 10.1088/1126-6708/2008/08/108.

[49] Axel Bredenstein, Ansgar Denner, Stefan Dittmaier, and Stefano Pozzorini. NLO QCD corrections to top anti-top bottom anti-bottom production at the LHC: 2. full hadronic results. *JHEP*, 03:021, 2010. doi: 10.1007/JHEP03(2010)021.

[50] Ruth Britto, Freddy Cachazo, and Bo Feng. Generalized unitarity and one-loop amplitudes in N = 4 super-Yang-Mills. *Nucl. Phys.*, B725: 275–305, 2005. doi: 10.1016/j.nuclphysb.2005.07.014.

[51] H. N. Brown et al. Precise measurement of the positive muon anomalous magnetic moment. *Phys. Rev. Lett.*, 86:2227–2231, 2001. doi: 10.1103/PhysRevLett.86.2227.

[52] L. M. Brown and R. P. Feynman. Radiative corrections to Compton scattering. *Phys. Rev.*, 85:231–244, 1952. doi: 10.1103/PhysRev.85.231.

[53] Manfred Böhm, Ansgar Denner, and Hans Joos. *Gauge Theories of the Strong and Electroweak Interaction*. Teubner Verlag, 2001. ISBN 9783519230458.

[54] John M. Campbell, E. W. Nigel Glover, and D. J. Miller. One-Loop Tensor Integrals in Dimensional Regularisation. *Nucl. Phys.*, B498: 397–442, 1997. doi: 10.1016/S0550-3213(97)00268-X.

[55] John M. Campbell, J. W. Huston, and W. J. Stirling. Hard Interactions of Quarks and Gluons: A Primer for LHC Physics. *Rept. Prog. Phys.*, 70:89, 2007. doi: 10.1088/0034-4885/70/1/R02.

[56] D. M. Capper, D. R. T. Jones, and P. van Nieuwenhuizen. Regularization by Dimensional Reduction of Supersymmetric and Non-supersymmetric Gauge Theories. *Nucl. Phys.*, B167:479, 1980. doi: 10.1016/0550-3213(80)90244-8.

[57] S. Catani, M. H. Seymour, and Z. Trocsanyi. Regularization scheme independence and unitarity in QCD cross sections. *Phys. Rev.*, D55: 6819–6829, 1997. doi: 10.1103/PhysRevD.55.6819.

[58] Stefano Catani, Stefan Dittmaier, and Zoltan Trocsanyi. One-loop singular behaviour of QCD and SUSY QCD amplitudes with massive partons. *Phys. Lett.*, B500:149–160, 2001. doi: 10.1016/S0370-2693(01)00065-X.

[59] G. M. Cicuta and E. Montaldi. Analytic renormalization via continuous space dimension. *Nuovo Cim. Lett.*, 4:329–332, 1972.

[60] John C. Collins. *Renormalization. An Introduction To Renormalization, The Renormalization Group, And The Operator Product Expansion*. Cambridge University Press, Cambridge, UK, 1984.

[61] John C. Collins, Davison E. Soper, and George F. Sterman. Factorization of Hard Processes in QCD. *Adv. Ser. Direct. High Energy Phys.*, 5:1–91, 1988.

[62] G. Cullen, J.-Ph. Guillet, G. Heinrich, T. Kleinschmidt, E. Pilon, T. Reiter, and M. Rodgers. Golem95C: A library for one-loop integrals with complex masses. arXiv:1101.5595 [hep-ph], 2011.

[63] Gavin Cullen, Maciej Koch-Janusz, and Thomas Reiter. spinney: A Form Library for Helicity Spinors. arXiv:1008.0803 [hep-ph], 2010.

[64] Gavin Cullen et al. Recent Progress in the Golem Project. *Nucl. Phys. Proc. Suppl.*, 205-206:67–73, 2010. doi: 10.1016/j.nuclphysbps.2010.08.021.

[65] M. Czakon. Automatized analytic continuation of Mellin-Barnes integrals. *Comput. Phys. Commun.*, 175:559–571, 2006. doi: 10.1016/j.cpc.2006.07.002.

[66] Henryk Czyz and Agnieszka Grzelinska. PHOKHARA 7.0 Monte Carlo generator: the narrow resonances implementation and new pion and kaon form factors. arXiv:1101.2967 [hep-ph], 2011.

[67] Henryk Czyz, Agnieszka Grzelinska, Johann H. Kuhn, and German Rodrigo. The radiative return at Phi- and B-factories: FSR for muon pair production at next-to-leading order. *Eur. Phys. J.*, C39:411–420, 2005. doi: 10.1140/epjc/s2004-02103-1.

[68] Henryk Czyz, Agnieszka Grzelinska, Johann H. Kuhn, and German Rodrigo. Electron positron annihilation into three pions and the radiative return. *Eur. Phys. J.*, C47:617–624, 2006. doi: 10.1140/epjc/s2006-02614-7.

[69] Andrei I. Davydychev. A Simple formula for reducing Feynman diagrams to scalar integrals. *Phys. Lett.*, B263:107–111, 1991. doi: 10.1016/0370-2693(91)91715-8.

[70] F. del Aguila and R. Pittau. Recursive numerical calculus of one-loop tensor integrals. *JHEP*, 07:017, 2004. doi: 10.1088/1126-6708/2004/07/017.

[71] A. Denner and S. Dittmaier. Scalar one-loop 4-point integrals. *Nucl. Phys.*, B844:199–242, 2011. doi: 10.1016/j.nuclphysb.2010.11.002.

[72] A. Denner, S. Dittmaier, S. Kallweit, and S. Pozzorini. NLO QCD corrections to WWbb production at hadron colliders. *Phys. Rev. Lett.*, 106:052001, 2011. doi: 10.1103/PhysRevLett.106.052001.

[73] Ansgar Denner. Techniques for calculation of electroweak radiative corrections at the one loop level and results for W physics at LEP-200. *Fortschr. Phys.*, 41:307–420, 1993.

[74] Ansgar Denner and S. Dittmaier. Reduction of one-loop tensor 5-point integrals. *Nucl. Phys.*, B658:175–202, 2003. doi: 10.1016/S0550-3213(03)00184-6.

[75] Ansgar Denner and S. Dittmaier. Reduction schemes for one-loop tensor integrals. *Nucl. Phys.*, B734:62–115, 2006. doi: 10.1016/j.nuclphysb.2005.11.007.

[76] Ansgar Denner, U. Nierste, and R. Scharf. A Compact expression for the scalar one loop four point function. *Nucl. Phys.*, B367:637–656, 1991. doi: 10.1016/0550-3213(91)90011-L.

[77] Ansgar Denner, S. Dittmaier, and T. Hahn. Radiative corrections to $ZZ \to ZZ$ in the electroweak standard model. *Phys. Rev.*, D56:117–134, 1997. doi: 10.1103/PhysRevD.56.117.

[78] Ansgar Denner, S. Dittmaier, M. Roth, and D. Wackeroth. Predictions for all processes $e^+ e^- \to 4$ fermions + gamma. *Nucl. Phys.*, B560:33–65, 1999. doi: 10.1016/S0550-3213(99)00437-X.

[79] Ansgar Denner, S. Dittmaier, M. Roth, and L. H. Wieders. Electroweak corrections to charged-current $e^+e^- \to 4$ fermion processes: Technical details and further results. *Nucl. Phys.*, B724:247–294, 2005. doi: 10.1016/j.nuclphysb.2005.06.033.

[80] Ganesh Devaraj and Robin G. Stuart. Reduction of one-loop tensor form-factors to scalar integrals: A general scheme. *Nucl. Phys.*, B519:483–513, 1998. doi: 10.1016/S0550-3213(98)00035-2.

[81] T. Diakonidis, J. Fleischer, J. Gluza, K. Kajda, T. Riemann, and J. B. Tausk. On the tensor reduction of one-loop pentagons and hexagons. *Nucl. Phys. Proc. Suppl.*, 183:109–115, 2008. doi: 10.1016/j.nuclphysbps.2008.09.091.

[82] T. Diakonidis, J. Fleischer, J. Gluza, K. Kajda, T. Riemann, and J. B. Tausk. A complete reduction of one-loop tensor 5- and 6-point integrals. *Phys. Rev.*, D80:036003, 2009. doi: 10.1103/PhysRevD.80.036003.

[83] T. Diakonidis, J. Fleischer, T. Riemann, and J. B. Tausk. A recursive reduction of tensor Feynman integrals. *Phys. Lett.*, B683:69–74, 2010. doi: 10.1016/j.physletb.2009.11.049.

[84] S. Dittmaier, P. Uwer, and S. Weinzierl. NLO QCD corrections to t anti-t + jet production at hadron colliders. *Phys. Rev. Lett.*, 98:262002, 2007. doi: 10.1103/PhysRevLett.98.262002.

[85] S. Dittmaier, P. Uwer, and S. Weinzierl. Hadronic top-quark pair production in association with a hard jet at next-to-leading order QCD: Phenomenological studies for the Tevatron and the LHC. *Eur. Phys. J.*, C59:625–646, 2009. doi: 10.1140/epjc/s10052-008-0816-y.

[86] P. Draggiotis, M. V. Garzelli, C. G. Papadopoulos, and R. Pittau. Feynman Rules for the Rational Part of the QCD 1-loop amplitudes. *JHEP*, 04:072, 2009. doi: 10.1088/1126-6708/2009/04/072.

[87] R. Keith Ellis and Giulia Zanderighi. Scalar one-loop integrals for QCD. *JHEP*, 02:002, 2008. doi: 10.1088/1126-6708/2008/02/002.

[88] R. Keith Ellis, W. T. Giele, and G. Zanderighi. Semi-numerical evaluation of one-loop corrections. *Phys. Rev.*, D73:014027, 2006. doi: 10.1103/PhysRevD.73.014027.

[89] R. Keith Ellis, Walter T. Giele, Zoltan Kunszt, and Kirill Melnikov. Masses, fermions and generalized D-dimensional unitarity. *Nucl. Phys.*, B822:270–282, 2009. doi: 10.1016/j.nuclphysb.2009.07.023.

[90] R. Keith Ellis, Kirill Melnikov, and Giulia Zanderighi. Generalized unitarity at work: First NLO QCD results for hadronic W+3 jet production. *JHEP*, 04:077, 2009. doi: 10.1088/1126-6708/2009/04/077.

[91] R. Keith Ellis, Kirill Melnikov, and Giulia Zanderighi. W+3 jet production at the Tevatron. *Phys. Rev.*, D80:094002, 2009. doi: 10.1103/PhysRevD.80.094002.

[92] R. Keith Ellis, Zoltan Kunszt, Kirill Melnikov, and Giulia Zanderighi. One-loop calculations in quantum field theory: from Feynman diagrams to unitarity cuts. arXiv:1105.4319 [hep-ph], 2011.

[93] L. D. Faddeev and V. N. Popov. Feynman diagrams for the Yang-Mills field. *Phys. Lett.*, B25:29–30, 1967. doi: 10.1016/0370-2693(67)90067-6.

[94] J. Fleischer and T. Riemann. A complete algebraic reduction of one-loop tensor Feynman integrals. *Phys. Rev.*, D83:073004, 2011. doi: 10.1103/PhysRevD.83.073004.

[95] J. Fleischer and M. Tentyukov. A Feynman diagram analyser DIANA: Graphic facilities. arXiv:hep-ph/0012189, 2000.

[96] J. Fleischer, F. Jegerlehner, and O. V. Tarasov. Algebraic reduction of one-loop Feynman graph amplitudes. *Nucl. Phys.*, B566:423–440, 2000. doi: 10.1016/S0550-3213(99)00678-1.

[97] J. Fleischer, F. Jegerlehner, and O. V. Tarasov. A new hypergeometric representation of one-loop scalar integrals in d dimensions. *Nucl. Phys.*, B672:303–328, 2003. doi: 10.1016/j.nuclphysb.2003.09.004.

[98] J. Fleischer, T. Riemann, and V. Yundin. One-loop tensor Feynman integral reduction with signed mincrs. 2011.

[99] Jochem Fleischer, Tord Riemann, and Valery Yundin. New results for algebraic tensor reduction of Feynman integrals. 2012.

[100] Rikkert Frederix, Stefano Frixione, Kirill Melnikov, and Giulia Zanderighi. NLO QCD corrections to five-jet production at LEP and the extraction of $\alpha_s(M_Z)$. *JHEP*, 11:050, 2010. doi: 10.1007/JHEP11(2010)050.

[101] H. M. Georgi, S. L. Glashow, M. E. Machacek, and Dimitri V. Nanopoulos. Charmed Particles From Two Gluon Annihilation In Proton Proton Collisions. *Ann. Phys.*, 114:273, 1978. doi: 10.1016/0003-4916(78)90270-1.

[102] W. Giele, E. W. Nigel Glover, and G. Zanderighi. Numerical evaluation of one-loop diagrams near exceptional momentum configurations. *Nucl. Phys. Proc. Suppl.*, 135:275–279, 2004. doi: 10.1016/j.nuclphysbps.2004.09.028.

[103] W. T. Giele and E. W. Nigel Glover. A calculational formalism for one-loop integrals. *JHEP*, 04:029, 2004. doi: 10.1088/1126-6708/2004/04/029.

[104] J. Gluza, K. Kajda, and T. Riemann. AMBRE - a Mathematica package for the construction of Mellin-Barnes representations for Feynman integrals. *Comput. Phys. Commun.*, 177:879–893, 2007. doi: 10.1016/j.cpc.2007.07.001.

[105] Janusz Gluza, Krzysztof Kajda, Tord Riemann, and Valery Yundin. New results for loop integrals: AMBRE, CSectors, hexagon. *POS*, ACAT08:124, 2008.

[106] Janusz Gluza, Krzysztof Kajda, Tord Riemann, and Valery Yundin. News on Ambre and CSectors. *Nucl. Phys. Proc. Suppl.*, 205-206:147–151, 2010. doi: 10.1016/j.nuclphysbps.2010.08.034.

[107] Janusz Gluza, Krzysztof Kajda, Tord Riemann, and Valery Yundin. Numerical Evaluation of Tensor Feynman Integrals in Euclidean Kinematics. *Eur. Phys. J.*, C71:1516, 2011. doi: 10.1140/epjc/s10052-010-1516-y.

[108] Nicolas Greiner, Alberto Guffanti, Thomas Reiter, and Jurgen Reuter. NLO QCD corrections to the production of two bottom-antibottom pairs at the LHC. arXiv:1105.3624 [hep-ph], 2011.

[109] J. F. Gunion and Z. Kunszt. Improved Analytic Techniques for Tree Graph Calculations and the G g q anti-q Lepton anti-Lepton Subprocess. *Phys. Lett.*, B161:333, 1985. doi: 10.1016/0370-2693(85)90774-9.

[110] T. Hahn and M. Perez-Victoria. Automatized one-loop calculations in four and D dimensions. *Comput. Phys. Commun.*, 118:153–165, 1999. doi: 10.1016/S0010-4655(98)00173-8.

[111] Thomas Hahn. Generating Feynman diagrams and amplitudes with FeynArts 3. *Comput. Phys. Commun.*, 140:418–431, 2001. doi: 10.1016/S0010-4655(01)00290-9.

[112] Thomas Hahn. Feynman Diagram Calculations with FeynArts, FormCalc, and LoopTools. *PoS*, ACAT2010:078, 2010.

[113] Thomas Hahn and Michael Rauch. News from FormCalc and LoopTools. *Nucl. Phys. Proc. Suppl.*, 157:236–240, 2006. doi: 10.1016/j.nuclphysbps.2006.03.026.

[114] Richard W Hamming. *Numerical methods for scientists and engineers (2nd ed.)*. Dover Publications, Inc., New York, NY, USA, 1986. ISBN 0-486-65241-6.

[115] S. W. Hawking. Particle Creation by Black Holes. *Commun. Math. Phys.*, 43:199–220, 1975. doi: 10.1007/BF02345020. [Erratum-ibid.46:206-206,1976].

[116] G. Heinrich, G. Ossola, T. Reiter, and F. Tramontano. Tensorial Reconstruction at the Integrand Level. *JHEP*, 10:105, 2010. doi: 10.1007/JHEP10(2010)105.

[117] Yozo Hida, Xiaoye S. Li, and David H. Bailey. http://crd.lbl.gov/~dhbailey/mpdist/, 2010.

[118] Valentin Hirschi et al. Automation of one-loop QCD corrections. *JHEP*, 05:044, 2011. doi: 10.1007/JHEP05(2011)044.

[119] Masataka Igarashi and Nobuya Nakazawa. QED RADIATIVE CORRECTIONS TO NEUTRINO COUNTING REACTION $e^+ e^- \to$ neutrino anti-neutrino gamma. *Nucl. Phys.*, B288:301, 1987. doi: 10.1016/0550-3213(87)90217-3.

[120] S. Jadach, B. F. L. Ward, and Z. Was. The precision Monte Carlo event generator KK for two-fermion final states in $e^+ e^-$ collisions. *Comput. Phys. Commun.*, 130:260–325, 2000. doi: 10.1016/S0010-4655(00)00048-5.

[121] S. Jadach, M. Melles, B. F. L. Ward, and S. A. Yost. Exact differential $O(\alpha^2)$ results for hard bremsstrahlung in $e^+ e^-$ annihilation to two fermions at and beyond LEP2 energies. *Phys. Rev.*, D65:073030, 2002. doi: 10.1103/PhysRevD.65.073030.

[122] Krzysztof Kajda, Tomas Sabonis, and Valery Yundin. QED Pentagon Contributions to $e^+e^- \to \mu^+\mu^-\gamma$. *Acta Phys. Pol.*, B40:3127–3133, 2009.

[123] T. Kinoshita. Mass singularities of Feynman amplitudes. *J. Math. Phys.*, 3:650–677, 1962.

[124] R. Kleiss and W. James Stirling. Spinor Techniques for Calculating p anti-p $\to W^\pm$ / Z^0 + Jets. *Nucl. Phys.*, B262:235–262, 1985. doi: 10.1016/0550-3213(85)90285-8.

[125] Johann H. Kuhn and German Rodrigo. The radiative return at small angles: Virtual corrections. *Eur. Phys. J.*, C25:215–222, 2002. doi: 10.1007/s10052-002-0999-6.

[126] Zoltan Kunszt, Adrian Signer, and Zoltan Trocsanyi. One loop helicity amplitudes for all $2 \to 2$ processes in QCD and N=1 supersymmetric Yang-Mills theory. *Nucl. Phys.*, B411:397–442, 1994. doi: 10.1016/0550-3213(94)90456-1.

[127] Zoltan Kunszt, Adrian Signer, and Zoltan Trocsanyi. Singular terms of helicity amplitudes at one loop in QCD and the soft limit of the

cross-sections of multiparton processes. *Nucl. Phys.*, B420:550–564, 1994. doi: 10.1016/0550-3213(94)90077-9.

[128] Wang-chuang Kuo, David Slaven, and Bing-Lin Young. Multi-Gluon Processes And The Slavnov-Taylor Identity. *Phys. Rev.*, D37:233, 1988. doi: 10.1103/PhysRevD.37.233.

[129] E. A. Kuraev and Victor S. Fadin. On Radiative Corrections to $e^+ e^-$ Single Photon Annihilation at High-Energy. *Sov. J. Nucl. Phys.*, 41: 466–472, 1985.

[130] T. D. Lee and M. Nauenberg. Degenerate Systems and Mass Singularities. *Phys. Rev.*, 133:B1549–B1562, 1964. doi: 10.1103/PhysRev.133.B1549.

[131] Fabio Maltoni and Tim Stelzer. MadEvent: Automatic event generation with MadGraph. *JHEP*, 02:027, 2003.

[132] Michelangelo L. Mangano and Stephen J. Parke. Multi-Parton Amplitudes in Gauge Theories. *Phys. Rept.*, 200:301–367, 1991. doi: 10.1016/0370-1573(91)90091-Y.

[133] Michelangelo L. Mangano, Mauro Moretti, Fulvio Piccinini, Roberto Pittau, and Antonio D. Polosa. ALPGEN, a generator for hard multiparton processes in hadronic collisions. *JHEP*, 07:001, 2003.

[134] P. Mastrolia, G. Ossola, T. Reiter, and F. Tramontano. Scattering AMplitudes from Unitarity-based Reduction Algorithm at the Integrand-level. *JHEP*, 08:080, 2010. doi: 10.1007/JHEP08(2010)080.

[135] Tom Melia, Kirill Melnikov, Raoul Rontsch, and Giulia Zanderighi. Next-to-leading order QCD predictions for W^+W^+jj production at the LHC. *JHEP*, 12:053, 2010. doi: 10.1007/JHEP12(2010)053.

[136] Tom Melia, Kirill Melnikov, Raoul Rontsch, and Giulia Zanderighi. NLO QCD corrections for W^+W^- pair production in association with two jets at hadron colliders. *Phys. Rev.*, D83:114043, 2011. doi: 10.1103/PhysRevD.83.114043.

[137] Kirill Melnikov and Markus Schulze. NLO QCD corrections to top quark pair production and decay at hadron colliders. *JHEP*, 08:049, 2009. doi: 10.1088/1126-6708/2009/08/049.

[138] Kirill Melnikov, Markus Schulze, and Andreas Scharf. QCD corrections to top quark pair production in association with a photon at hadron colliders. *Phys. Rev.*, D83:074013, 2011. doi: 10.1103/PhysRevD.83.074013.

[139] D. B. Melrose. Reduction of Feynman diagrams. *Nuovo Cim.*, 40: 181–213, 1965.

[140] C. Milardi et al. Crab waist collision at DAFNE. *ICFA Beam Dyn. Newslett.*, 48:23–33, 2009.

[141] K. Nakamura et al. Review of particle physics. *J. Phys.*, G37:075021, 2010. doi: 10.1088/0954-3899/37/7A/075021.

[142] Dao Thi Nhung and Le Duc Ninh. D0C : A code to calculate scalar one-loop four-point integrals with complex masses. *Comput. Phys. Commun.*, 180:2258–2267, 2009. doi: 10.1016/j.cpc.2009.07.012.

[143] Paulo Nogueira. Automatic Feynman graph generation. *J. Comput. Phys.*, 105:279–289, 1993. doi: 10.1006/jcph.1993.1074.

[144] Giovanni Ossola, Costas G. Papadopoulos, and Roberto Pittau. Reducing full one-loop amplitudes to scalar integrals at the integrand level. *Nucl. Phys.*, B763:147–169, 2007. doi: 10.1016/j.nuclphysb.2006.11.012.

[145] Giovanni Ossola, Costas G. Papadopoulos, and Roberto Pittau. Cut-Tools: a program implementing the OPP reduction method to compute one-loop amplitudes. *JHEP*, 03:042, 2008. doi: 10.1088/1126-6708/2008/03/042.

[146] G. Passarino and M. J. G. Veltman. One Loop Corrections for $e^+ e^-$ Annihilation Into $\mu^+ \mu^-$ in the Weinberg Model. *Nucl. Phys.*, B160:151, 1979. doi: 10.1016/0550-3213(79)90234-7.

[147] Michael E. Peskin and Dan V. Schroeder. *An Introduction To Quantum Field Theory*. Westview Press, 1995. ISBN 9780201503975.

[148] A. Pukhov et al. CompHEP: A package for evaluation of Feynman diagrams and integration over multi-particle phase space. User's manual for version 33. arXiv:hep-ph/9908288, 1999.

[149] Thomas Reiter. Optimising Code Generation with haggies. *Comput. Phys. Commun.*, 181:1301–1331, 2010. doi: 10.1016/j.cpc.2010.01.012.

[150] German Rodrigo, Henryk Czyz, Johann H. Kuhn, and Marcin Szopa. Radiative return at NLO and the measurement of the hadronic cross-section in electron positron annihilation. *Eur. Phys. J.*, C24:71–82, 2002. doi: 10.1007/s100520200912.

[151] Daniel Shanks. Non-linear transformations of divergent and slowly convergent sequences. *J. Math. Phys.*, 34:1–42, 1955.

[152] Warren Siegel. Supersymmetric Dimensional Regularization via Dimensional Reduction. *Phys. Lett.*, B84:193, 1979. doi: 10.1016/0370-2693(79)90282-X.

[153] Adrian Signer and Dominik Stockinger. Factorization and regularization by dimensional reduction. *Phys. Lett.*, B626:127–138, 2005. doi: 10.1016/j.physletb.2005.08.112.

[154] Adrian Signer and Dominik Stockinger. Dimensional Reduction and Hadronic Processes. *AIP Conf. Proc.*, 1078:256–258, 2009. doi: 10.1063/1.3051926.

[155] Adrian Signer and Dominik Stockinger. Using Dimensional Reduction for Hadronic Collisions. *Nucl. Phys.*, B808:88–120, 2009. doi: 10.1016/j.nuclphysb.2008.09.016.

[156] Vladimir A. Smirnov. *Feynman Integral Calculus*. Springer, Berlin, 2006.

[157] Mark Srednicki. *Quantum Field Theory*. Cambridge University Press, 2007. ISBN 9780521864497.

[158] Robin G. Stuart. Algebraic Reduction Of One Loop Feynman Diagrams To Scalar Integrals. *Comput. Phys. Commun.*, 48:367–389, 1988. doi: 10.1016/0010-4655(88)90202-0.

[159] Robin G. Stuart and A. Gongora. Algebraic Reduction Of One Loop Feynman Diagrams To Scalar Integrals. 2. *Comput. Phys. Commun.*, 56:337–350, 1990. doi: 10.1016/0010-4655(90)90019-W.

[160] Gerard 't Hooft and M. J. G. Veltman. Combinatorics of gauge fields. *Nucl. Phys.*, B50:318–353, 1972. doi: 10.1016/S0550-3213(72)80021-X.

[161] Gerard 't Hooft and M. J. G. Veltman. Regularization and Renormalization of Gauge Fields. *Nucl. Phys.*, B44:189–213, 1972. doi: 10.1016/0550-3213(72)90279-9.

[162] Gerard 't Hooft and M. J. G. Veltman. Scalar One Loop Integrals. *Nucl. Phys.*, B153:365–401, 1979. doi: 10.1016/0550-3213(79)90605-9.

[163] O. V. Tarasov. Connection between Feynman integrals having different values of the space-time dimension. *Phys. Rev.*, D54:6479–6490, 1996. doi: 10.1103/PhysRevD.54.6479.

[164] M. Tentyukov and J. Fleischer. DIANA, a program for Feynman diagram evaluation. arXiv:hep-ph/9905560, 1999.

[165] M. Tentyukov and J. Fleischer. A Feynman diagram analyser DIANA. *Comput. Phys. Commun.*, 132:124–141, 2000. doi: 10.1016/S0010-4655(00)00147-8.

[166] M. Tentyukov and J. Fleischer. A Feynman diagram analyzer DIANA: Recent development. *Nucl. Instrum. Meth.*, A502:570–572, 2003. doi: 10.1016/S0168-9002(03)00505-9.

[167] A. van Hameren. OneLOop: for the evaluation of one-loop scalar functions. arXiv:1007.4716 [hep-ph], 2010.

[168] Andre van Hameren, Jens Vollinga, and Stefan Weinzierl. Automated computation of one-loop integrals in massless theories. *Eur. Phys. J.*, C41:361–375, 2005. doi: 10.1140/epjc/s2005-02229-6.

[169] W. L. van Neerven and J. A. M. Vermaseren. LARGE LOOP INTEGRALS. *Phys. Lett.*, B137:241, 1984. doi: 10.1016/0370-2693(84)90237-5.

[170] G. J. van Oldenborgh. FF: A Package to evaluate one loop Feynman diagrams. *Comput. Phys. Commun.*, 66:1–15, 1991. doi: 10.1016/0010-4655(91)90002-3.

[171] G. J. van Oldenborgh and J. A. M. Vermaseren. New Algorithms for One Loop Integrals. *Z. Phys.*, C46:425–438, 1990. doi: 10.1007/BF01621031.

[172] J. A. M. Vermaseren. New features of FORM. arXiv:math-ph/0010025, 2000.

[173] Steven Weinberg. A Model of Leptons. *Phys. Rev. Lett.*, 19:1264–1266, 1967. doi: 10.1103/PhysRevLett.19.1264.

[174] Ernst Joachim Weniger. Nonlinear sequence transformations for the acceleration of convergence and the summation of divergent series. *Computer Physics Reports*, 10(5-6):189–371, 1989. doi: 10.1016/0167-7977(89)90011-7.

[175] Kenneth G. Wilson. Quantum field theory models in less than four-dimensions. *Phys. Rev.*, D7:2911–2926, 1973. doi: 10.1103/PhysRevD.7.2911.

[176] P. Wynn. On a device for computing the $e_m(S_n)$ transformation. *Math. Tables Aids Comput.*, 10(54):91–96, 1956.

[177] D. R. Yennie, Steven C. Frautschi, and H. Suura. The infrared divergence phenomena and high-energy processes. *Ann. Phys.*, 13:379–452, 1961. doi: 10.1016/0003-4916(61)90151-8.

List of Figures

2.1. Tree diagrams for $2 \to 2$ scattering in ϕ^3 8

3.1. All-outgoing momenta labeling 20
3.2. Basis integrals for tensor pentagons reduction 35

4.1. Rank 3 pentagon form-factors calculation flowchart 44
4.2. All-incoming momenta labeling 49
4.3. Passarino-Veltman reduction and PJFry code in small Gram region 56
4.4. Small Gram expansion accuracy for E_{3333} 58
4.5. Passarino-Veltman reduction accuracy in small Gram region . 59
4.6. Massless kinematics contraction accuracy $\delta_{\text{cut}} = 10^{-11}$ 64
4.7. Massless kinematics contraction accuracy $\delta_{\text{cut}} = 10^{-3}$ 65
4.8. Massive pentagons in $t\bar{t} + j$ 66
4.9. Massive kinematics contraction accuracy $E = 2.2m$ 67
4.10. Massive kinematics contraction accuracy $E = 3m$ part 1 . . . 68
4.11. Massive kinematics contraction accuracy $E = 4m$ 69
4.12. Five gluon helicity amplitude accuracy for $\delta_{\text{cut}} = 10^{-7}$ 70
4.13. Five gluon helicity amplitude accuracy for $\delta = 10^{-3}$ 70

5.1. Tree diagrams for $e^+e^- \to \mu^+\mu^-\gamma$ 75
5.2. Muon pair invariant mass distribution for KLOE 85
5.3. Angular distributions of μ^+ and μ^- for KLOE 85
5.4. Muon pair invariant mass distribution for BaBar 86
5.5. Angular distributions of μ^+ and μ^- for BaBar 86
5.6. Angular distribution of photon for BaBar 87

5.7. "Penta" contribution to muon pair invariant mass distribution for KLOE 88
5.8. "Penta" contribution to angular distributions of μ^+ and μ^- for KLOE 88
5.9. "Penta" contribution to forward-backward asymmetry of μ^+ for KLOE 88
5.10. "Penta" contribution to muon pair invariant mass distribution for BaBar 89
5.11. "Penta" contribution to angular distributions of μ^+ and μ^- for BaBar 89
5.12. "Penta" contribution to forward-backward asymmetry of μ^+ for BaBar 89
5.13. "Penta" contribution to angular distribution of photon for BaBar 90
5.14. Relative contributions to Q^2 distribution for KLOE 90
5.15. Relative size of different contributions to the Q^2 distribution normalized on Born for BaBar setup (left σ, right $\bar{\sigma}$). 90
5.16. Relative contributions to angular distributions of μ^+ and μ^- for KLOE 91
5.17. Relative contributions to angular distributions of μ^+ and μ^- for KLOE 91
5.18. Relative contributions to forward-backward asymmetry of μ^+ for KLOE 92
5.19. Relative contributions to forward-backward asymmetry of μ^+ for KLOE 92

D.1. Bubble topology diagrams for $e^+e^- \to \mu^+\mu^-\gamma$ 106
D.2. Triangle topology diagrams for $e^+e^- \to \mu^+\mu^-\gamma$ 107
D.3. Box topology diagrams for $e^+e^- \to \mu^+\mu^-\gamma$ 108
D.4. Pentagon topology diagrams for $e^+e^- \to \mu^+\mu^-\gamma$ 109

List of Tables

4.1. Tensor form-factor functions' names 50

5.1. Classes of diagrams for $e^+e^- \to \mu^+\mu^-\gamma$ process 77

5.2. Phase-space cuts for KLOE and BaBar settings. Q^2 is the invariant mass squared of the muon pair. 84

Acknowledgements

First and foremost I wish to thank my adviser Dr. Tord Riemann for his advice and guidance during the course of my PhD studies. His ideas set the main directions of this thesis while leaving enough flexibility and freedom for independent work.

I would like to thank committee members Prof. Dr. Jan Plefka, Prof. Dr. Peter Uwer, Prof. Dr. Henryk Czyż, Prof. Dr. Tomas Lohse and Prof. Dr. Michael Müller-Preußker for useful feedback on the thesis and interesting questions during my thesis defense.

During my years in DESY I have enjoyed the friendly atmosphere and numerous coffee-room conversations with members of theory and NIC groups. In particular I would like thank Dr. Simon Badger and Dr. Simone Alioli for many helpful discussions on QCD-related topics and other things.

I thank the University of Silesia theory group for hospitality during my stay there. It was a pleasure to work with Prof. Dr. Janusz Gluza and Dr. Krzysztof Kajda.

I owe my deepest gratitude to my friends for their support. I am grateful to Dirk Hesse for being able to write these lines even after our great (and sometimes thrilling) climbing adventures together.

This thesis would not have been possible without enlightening bus stop conversation with Misha Lemeshko. I thank him for that and for numerous friendly occasions of "nu nelzia zhe tak pit" during which we discussed questions of life the universe and everything.

Finally, I thank my family for their love and support.

i want morebooks!

Buy your books fast and straightforward online - at one of world's fastest growing online book stores! Environmentally sound due to Print-on-Demand technologies.

Buy your books online at
www.get-morebooks.com

Kaufen Sie Ihre Bücher schnell und unkompliziert online – auf einer der am schnellsten wachsenden Buchhandelsplattformen weltweit! Dank Print-On-Demand umwelt- und ressourcenschonend produziert.

Bücher schneller online kaufen
www.morebooks.de

VDM Verlagsservicegesellschaft mbH
Heinrich-Böcking-Str. 6-8   Telefon: +49 681 3720 174   info@vdm-vsg.de
D - 66121 Saarbrücken   Telefax: +49 681 3720 1749   www.vdm-vsg.de

Printed by Books on Demand GmbH, Norderstedt / Germany